HIGH-RESOLUTION MASS SPECTROMETRY AND ITS DIVERSE APPLICATIONS

Cutting-Edge Techniques and Instrumentation

HIGH-RESOLUTION MASS SPECTROMETRY AND ITS DIVERSE APPLICATIONS

Cutting-Edge Techniques and Instrumentation

Edited by

Sreeraj Gopi, PhD

Sabu Thomas, PhD

Augustine Amalraj, PhD

Shintu Jude, MSc

AAP | APPLE ACADEMIC PRESS

First edition published 2023

Apple Academic Press Inc.
1265 Goldenrod Circle, NE,
Palm Bay, FL 32905 USA

760 Laurentian Drive, Unit 19,
Burlington, ON L7N 0A4, CANADA

CRC Press
6000 Broken Sound Parkway NW,
Suite 300, Boca Raton, FL 33487-2742 USA

4 Park Square, Milton Park,
Abingdon, Oxon, OX14 4RN UK

Apple Academic Press exclusively co-publishes with CRC Press, an imprint of Taylor & Francis Group, LLC

Library and Archives Canada Cataloguing in Publication

Title: High-resolution mass spectrometry and its diverse applications : cutting-edge techniques and instrumentation / edited by Sreeraj Gopi, PhD, Sabu Thomas, PhD, Augustine Amalraj, PhD, Shintu Jude, MSc.
Names: Gopi, Sreeraj, editor. | Thomas, Sabu, editor. | Amalraj, Augustine, editor. | Jude, Shintu, editor.
Description: First edition. | Includes bibliographical references and index.
Identifiers: Canadiana (print) 20220281351 | Canadiana (ebook) 20220281386 | ISBN 9781774911129 (hardcover) | ISBN 9781774911136 (softcover) | ISBN 9781003304258 (ebook)
Subjects: LCSH: High resolution spectroscopy.
Classification: LCC QC454.H618 H54 2023 | DDC 535.8/4—dc23

Library of Congress Cataloging-in-Publication Data

CIP data on file with US Library of Congress

ISBN: 978-1-77491-112-9 (hbk)
ISBN: 978-1-77491-113-6 (pbk)
ISBN: 978-1-00330-425-8 (ebk)

About the Editors

Sreeraj Gopi, PhD
Plant Lipids Private Limited, Kerala, India

Sreeraj Gopi, PhD, is an industrial scientist with a doctorate in organic chemistry, nanotechnology, and nanodrug delivery. He has been working in the area of natural products, isolation, and establishing biological activities. He has published more than 100 international articles and has filed more than 75 international patents. Dr. Gopi is a fellow of the Royal Society of Chemistry and a fellow of the American College of Nutrition, USA. Dr. Gopi is also co-author of many books and is also an invited speaker at many universities. Dr. Gopi is serving as an adjunct professor at many universities, including Stockholm University, Sweden; Siberian Federal University, Russia; Mahatma Gandhi University, India; and Gdansk University of Technology, Poland. He has traveled to more than 35 countries to give lectures.

Sabu Thomas, PhD
Vice-Chancellor, Mahatma Gandhi University;
Founder Director and Professor, International and Inter University Centre for Nanoscience and Nanotechnology, Kottayam, India

Sabu Thomas, PhD, is a Professor of Polymer Science and Engineering, Director of the Center for Nanoscience and Nanotechnology, and Vice-Chancellor of Mahatma Gandhi University, Kottayam, India. Prof. Thomas has substantial research experience across nanoscience, polymer science and engineering, polymer nanocomposites, elastomers, polymer blends, interpenetrating polymer networks, polymer membranes, green composites, and nanocomposites, nanomedicine, and green nanotechnology. He has received many national and international awards and has published over 1,000 peer-reviewed research papers. He has co-edited 135 books and has 15 patents.

Augustine Amalraj, PhD
Deputy Manager, Department of Research and Development, Plant Lipids Private Limited, Cochin, India

Augustine Amalraj, PhD, is currently working as a Deputy Manager in the Department of Research and Development in Plant Lipids Private Limited, Cochin, India. He obtained his doctoral degree in chemistry from Gandhigram Rural Institute – Deemed University, Gandhigram, Tamil Nadu, India. His research interests are in applied chemistry, food chemistry, natural product chemistry, environmental chemistry, chemosensors, and polymer and nanocomposite materials. He has published more than 60 research articles in international journals as well as 15 book chapters.

Shintu Jude, MSc
Assistant Manager, Research and Development Department, Plant Lipids Private Limited, Cochin, India

Shintu Jude, MSc, is working as Assistant Manager in Research and Development department at Plant Lipids Private Limited, Cochin, India. Mrs. Jude has completed her postgraduation at Mahatma Gandhi University, India. She is working on mass spectrometric instruments for natural products and metabolites. She has published more than 20 research articles and 10 book chapters.

Contents

Contributors

Augustine Amalraj
Aurea Biolabs (P) Ltd., R&D Center, Kolenchery, Cochin – 682311, Ernakulam, Kerala, India, E-mail: augustin14amal@gmail.com

E. Jackcina Stobel Christy
Department of Chemistry, The Gandhigram Rural Institute – Deemed to be University, Gandhigram, Dindigul – 624302, Tamil Nadu, India

Xueming Dong
College of Chemical and Biological Engineering, Shandong University of Science and Technology, Qingdao, Shandong – 266590, China; Department of Chemistry, Purdue University, Brown Building, 560 Oval Drive, West Lafayette, Indiana – 47907, United States

Xing Fan
College of Chemical and Biological Engineering, Shandong University of Science and Technology, Qingdao, Shandong – 266590, China; Key Laboratory of Coal Clean Conversion and Chemical Engineering Process, Xinjiang Uygur Autonomous Region, College of Chemistry and Chemical Engineering, Xinjiang University, Urumqi – 830046, China, E-mail: fanxing@cumt.edu.cn

Sreeraj Gopi
Chemical Faculty, Department of Chemistry, Gdansk University of Technology, Gdańsk, Poland; R&D Center, Aurea Biolabs (P) Ltd., Kolenchery, Cochin – 682311, Ernakulam, Kerala, India, E-mail: sreerajgopi@yahoo.com

Józef T. Haponiuk
Chemical Faculty, Department of Chemistry, Gdansk University of Technology, Gdańsk, Poland

Joby Jacob
R&D Center, Aurea Biolabs (P) Ltd., Kolenchery, Cochin – 682311, Ernakulam, Kerala, India, E-mail: jobyjacob799@gmail.com

Shintu Jude
R&D Center, Aurea Biolabs (P) Ltd., Kolenchery, Cochin – 682311, Ernakulam, Kerala, India, E-mail: shintujude2@gmail.com

Akhila Nair
Department of Chemistry, Gdansk University of Technology, Gdansk, Poland; R&D Center, Aurea Biolabs (P) Ltd., Kolenchery, Cochin – 682311, Ernakulam, Kerala, India

Reshma V. R. Nair
R&D Center, Aurea Biolabs (P) Ltd., Kolenchery, Cochin – 682311, Ernakulam, Kerala, India, E-mail: reshmavr1992@gmail.com

Anitha Pius
Department of Chemistry, The Gandhigram Rural Institute – Deemed to be University, Gandhigram, Dindigul – 624302, Tamil Nadu, India, Phone: +91-451-2452371 (O), Fax: +91-451-2454466 (O), E-mail: dranithapius@gmail.com

A. Rajeswari
Department of Chemistry, The Gandhigram Rural Institute – DU, Gandhigram, Dindigul – 624302, Tamil Nadu, India

S. Subasini
Department of Chemistry, The Gandhigram Rural Institute – Deemed to be University, Gandhigram, Dindigul – 624302, Tamil Nadu, India

Karthik Varma
R&D Center, Aurea Biolabs (P) Ltd., Kolenchery, Cochin – 682311, Ernakulam, Kerala, India, E-mail: ackvarmaapm@gmail.com

Abbreviations

3D	three dimensional
AC-SWV	alternating current square wave voltage
AEOs	alcohol polyethoxylates
AFM	atomic force microscopy
AMP	ampicillin
ANFO	ammonium nitrate fuel oil
AOAC	Association of Official Analytical Chemists
AP	amphetamine
AP	atmospheric pressure
APCI	atmospheric pressure chemical ionization
APEOs	alkylphenol polyethoxylates
API	American Petroleum Institute
APPI	atmospheric pressure photoionization
ASAP	ambient solid analysis probe
ATM	automated teller machine
BCME	bis-chloromethyl ether
BD	bligh-dyer
BE	benzoylecgonine
BFRs	brominated flame retardants
bNAbs	broadly neutralizing antibodies
BSC	bocaiuva (*Acrocomia aculeata*) seed cake
CBD	cannabidiol
CE	capillary electrophoresis
CESI	capillary ESI
CID	collision-induced dissociation
CNS	central nervous system
COC	cocaine
CPs	chlorinated paraffins
CSF	cerebrospinal fluid
CT	computerized tomography
CTPM	Chinese traditional patent medicines
CUP	current-use pesticides

DART	direct analysis in real-time
DART-HRMS	direct analysis in real-time coupled to high-resolution mass spectrometry
DART-MS	direct analysis in real-time mass spectrometry
DART-TOFMS	DART-time-of-flight-MS
DBE	double bond equivalence
DDA	data-dependent acquisition
DDT	dichloro-diphenyl-trichloroethane
DESI	desorption electrospray ionization
DFC	drug-facilitated crimes
DFSA	drug-facilitated sexual assault
DIA	data-independent acquisition
DI-nESI-HRMS	direct infusion nanoelectrospray high-resolution mass spectrometry
DLLME	dispersive liquid-liquid microextraction
DM	drug metabolism
DP	dissolved phase
dSPE	dispersive SPE
DUID	driving under the influence of drugs
DVFX	O-desmethylvenlafaxine
EDTA	ethylene diamine tetraacetate
EI	electron ionization
EM	electron microscopy
ESI	electron spray ionization
FAME	fatty acid methyl ester
FBF	find by formula
FCPs	fluorine-containing pesticides
FD	field desorption
FI-TOF-MS	flow injection-quadrupole-time-of-flight mass spectrometry
FO	Folch
FPD	flame photometric detection
FT Orbitrap	Fourier transform orbitrap
FTICR	Fourier transform ion cyclotron resonance
FT-ICR-MS	Fourier transform-ion cyclotron resonance-mass spectrometer
FTIR	Fourier transform infrared spectroscopy
FWHM	full width at half maximum

FXC	Fufang Xueshuantong capsule
GABA	gamma-aminobutyric acid
GC	gas chromatography
GCFPD	gas chromatography-flame photometric detection
GCIB	gas cluster ion beams
GCIB-SIMS	gas cluster ion beam secondary ion mass spectrometry
GC–MS	GC-mass spectrometer
GSR	gunshot residues
H/D	hydrogen/deuterium
HA	atrazine-2- hydroxy
HCA	hierarchical cluster analysis
HCB	hexachlorobenzene
HCHs	hexachlorocyclohexanes
HESI	heated electrospray ionization
HFDI-MS	hydrogen flame desorption ionization mass spectrometry
HGP	human genome project
HMT	hexamethylenetriamine
HPLC	high-performance liquid chromatography
HR	high resolution
HR-FS	high-resolution full scan
HRMS	high-resolution mass spectrometry
HRMS-QTOF	high-resolution mass spectra-quadrupole time of flight
HR-TOF-MS	high-resolution time of flight mass spectrometry
I-CIMS	chemical ionization mass spectrometer using iodide
IC-PAD	ion chromatography coupled with pulsed amperometric detection
ICP-MS	inductively coupled plasma mass spectrometry
ID	identification
IEC	internal electrode capillary
IEDs	improvised explosive devices
IMS	ion mobility spectrometry
INP	Itatiaia National Park

IR	infrared spectroscopy
IS	internal standard
ISCAD	ion-source collision-activated dissociation
IT	ion-trap
ITMS	ion trap MS
KET	ketamine
LA	laser ablation
LAESI	laser ablation ESI
LAESI-MS	laser ablation electrospray ionization mass spectrometry
LC	liquid chromatography
LC-DAD-Orbitrap MS	liquid chromatography diode array detector-orbitrap mass spectrometry
LC-MS	liquid chromatography-tandem mass spectrometry
LC-MS/MS	LC-tandem mass spectrometer
LDA	linear discriminant analysis
LDPE	low-density polyethylene
LIAD	laser-induced acoustic desorption
LLE	liquid-liquid extraction
LLOQ	lower limit of quantification
LOD	limit of detection
LOQ	limit of quantitation
LRMS	low resolution mass spectrometry
LSD	lysergic acid diethylamide
MA	mass accuracy
MA	Matyash
MA	methamphetamine
MALDI	matrix-assisted laser desorption ionization
MALDI-MSI	matrix-assisted laser desorption/ionization-mass spectrometry imaging
MALDI-TOFMS	matrix-assisted laser desorption ionization-time of flight mass spectrometry
MBTOF	multibounce time of flight
MDA	3,4-methylenedioxyamphetamine
mDa	milli Dalton
MDMA	3,4-methylenedioxymethamphetamine
MDPV	methylenedioxypyrovalerone

MEC	methylethcathione
MEMS	micro electro mechanical system
miRNAs	microRNAs
MMC	methylmethcathione
MP	micropollutants
MRI	magnetic resonance imaging
MRLs	maximum residue limits
MRM	multiple reaction monitoring
MRMs	multi-residue methods
MS	mass spectrometry
MSAD	multipole storage assisted dissociation
MSI	mass spectrometry imaging
MSPD	matrix solid-phase dispersion
MVOC	microbial volatile organic compounds
NEMS	nanoelectromechanical system
NG	nitroglycerin
NK	norketamine
NM2AI	N-methyl-2-aminoindane
NMR	nuclear magnetic resonance
NNIs	neonicotinoid insecticides
NPD	new psychoactive drugs
NTM	nontuberculous mycobacteria
NTS	non-target screening
OCP	organochlorine pesticides
OF	oral fluid
OPPs	organophosphate pesticides
PBDEs	polybrominated diphenyl ethers
PCA	principal component analysis
PCBs	polychlorinated biphenyls
PCDDs	polychlorinated dibenzo-p-dioxins
PCR	polymerase chain reaction
PCs	principal components
PD	pharmacodynamics
PDMS	polydimethylsiloxane
PESI	probe ESI
PFCs	perfluorinated compounds
PK	pharmacokinetic
POCIS	polar organic chemical integrative sampler

PP	particulate phase
ppm	parts per million
PTMs	post-translational modifications
PTR	proton transfer reaction
QA	quality assurance
QC	quality control
QqQLIT	triple quadrupole linear ion trap
QTOF	quadrupole time of flight
QTTs	quercotriterpenosides
RDBE	ring and double bond equivalent
RH-GC	resistive heating gas chromatography
RMS	root-mean-square
RSD	relative standard deviation
RT	retention time
S/N	signal to noise ratio
SALLE	salting-out liquid-liquid extraction
SARA	saturate, aromatic, resin, and asphaltene
SECTOR	sector mass spectrometer
SFC	supercritical fluid chromatography
SHW	sewage hospital water
SIM	selected ion monitoring
SIMS	secondary ion mass spectrometry
SIR	selected ion recording
SMC	systematic multi-component
SMPS	scanning mobility particle sizer
SOM	soil organic matter
SONP	Serra Dos Órgãos National Park
SPE	solid-phase extraction
SPI-ESI	synchronized polarization induced electrospray ionization
SPME	solid-phase microextraction
SRB	sulforhodamine B
SRM	selected reaction monitoring
SSRI	selective serotonin reuptake inhibitor
SWATH	sequential window acquisition of all theoretical fragment ion spectra
SWGDRUG	scientific working group for the analysis of seized drugs

SXT	Shuxiong Tablet
TCM	traditional Chinese medicine
TD	therapeutic drug
TD-DART-HRMS	thermal desorption–direct analysis in real-time–high-resolution mass spectrometry
TDM	therapeutic drug monitoring
TET	tetracycline
THC	Δ^9-tetrahydrocannabinol
THCA	11-carboxy-delta-9-tetrahydrocannabinol
TI	triterpenoids index
TLC	thin layer chromatography
ToF	time of flight
TOF-SIMS	time of flight secondary ion mass spectrometry
TP	transformation products
UHPLC	ultra-high performance liquid chromatography
UPLCQ	ultra-performance liquid chromatography quadrupole
USFDA	The United States Food and Drug Administration
UTRs	3'-untranslated regions
Vas	veterinary antibiotics
VFX	venlafaxine
VP	ventriculoperitoneal
VUVDI	vacuum ultraviolet laser desorption/ionization
YCHD	Yinchenhao decoction

Preface

The story of mass spectrometry (MS) begins with the invention of aids for mass determination. With the introduction of high-resolution mass spectrometers, some magic spells have been added to it. High resolution (HR) MS has changed the nature of experiments and investigation of so many analytical realms. Exact mass determination, HR, specificity – yes, through each and every special feature provided by HRMS instruments, it is now possible to determine the composition of the analyte of interest, both qualitatively and quantitatively. The technical evolution happened to allow HRMS instrumentation to change all the conventional areas of analysis at a great pace. Along with the untargeted workflows, HRMS has also been proven as a promising platform for quantitative applications, even for larger molecules, which is considered to be a hurdle in the field.

This book presents a wide range of knowledge of the technologies and applications of HRMS in different areas of analysis.

Basic knowledge of the technique and instruments makes the path easy to move further. Hence, the book begins with the basic instrumentation techniques and provides a wider path to move on to. HRMS plays an important role in structure elucidation and unknown determination in many fields. Indeed, it is a great measure to be used for quantitative analyzes. Together, these properties make the technique a strong aid in clinical laboratories, drug development, quality management systems, forensic analyses, pesticide evaluation, environmental impact assessment, fuel analysis, etc. The samples analyzed in HRMS can range from single cells to objects beyond the globe. We have tried to include all those facts in the book.

The editors (Dr. Sreeraj Gopi, Dr. A. Amalraj, and Ms. Shintu Jude) would like to take this chance to express thanks to the colleagues, staff, and management of Plant Lipids Private Limited for their help and enormous support.

We hope the book can be a guide to your way to MS and a better companion for your analytical strides.

CHAPTER 1

HRMS: Fundamental Concepts, and Instrumentation

KARTHIK VARMA,[1] SREERAJ GOPI,[2] and JÓZEF T. HAPONIUK[2]

[1]R&D Center, Aurea Biolabs (P) Ltd., Kolenchery, Cochin – 682311, Ernakulam, Kerala, India, E-mail: ackvarmaapm@gmail.com

[2]Chemical Faculty, Gdansk University of Technology, Gdańsk, Poland, E-mail: sreerajgopi@yahoo.com (S. Gopi)

ABSTRACT

Scientific communities always depend on high-end sophisticated instruments for the development of new molecules and metabolites in the research field. With the introduction of high-resolution mass spectrometry (HRMS) in the analytical industry, major hurdles in the field of detection of the bioactive and their metabolite even in a lower concentration level can be achieved. HRMS can be highly effectively used in various fields like the natural products industry, clinical, and forensic chemistry, and many other valuable biomedical applications. Through different ionization and data acquisition methods, sensitive, and reliable results can be achieved using HRMS. The chapter aims to present an overview of the fundamentals, concepts, and instrumentation techniques of HRMS.

High-Resolution Mass Spectrometry and Its Diverse Applications: Cutting-Edge Techniques and Instrumentation. Sreeraj Gopi, PhD, Sabu Thomas, PhD, Augustine Amalraj, PhD, & Shintu Jude, MSc (Editors)

1.1 INTRODUCTION

The past few decades have witnessed the use of high-end sophisticated instruments for the quantification and qualification of unknown molecules in the scientific community. The use of such instruments has paved the way opening to researchers and industrialists to explore and concentrate more on the use of sophisticated instruments. The use of reliable and sensitive instruments will surely unlock the mystery of unknown molecules and concentrate more on developmental activities.

The fundamental concept involved in the mass spectrometric analysis is the determination of molecules, which sensitively detect molecules through their ions and allows to get the molecular information in different dimensions according to the method applied (Petras et al., 2017). The mass spectrometry (MS) does not directly measure the mass, instead, it measures the "m/z" in which 'm' represents the relative mass of an ion and 'z' is the charge of the ion. High-resolution mass spectra-quadrupole time of flight (HRMS-QTOF) has been one of the promising innovations in the field of high-resolution mass spectral analysis. HRMS is used in various fields like medical, clinical, and forensic toxicology, food safety, environmental protection, bioengineering, and chemical engineering due to its wide mass range, fast scanning speed, high degree of sensitivity (full scan mode), high resolution (HR), and precise molecular weight measurement. The different types of HRMS include time of flight (TOF), sector mass spectrometer (SECTOR), Fourier transform-ion cyclotron resonance-mass spectrometer (FT-ICR-MS), and Fourier transform orbitrap (FT Orbitrap) (Hu et al., 2005).

The chapter aims to present an overview of the fundamental terms involved in HRMS, instrumentation techniques, molecular ion determination, different sampling procedures, etc.

1.2 BASIC CONCEPTS AND INSTRUMENTATION TECHNIQUES

1.2.1 LOW-RESOLUTION MASS SPECTROMETRY (LRMS) VERSUS HIGH-RESOLUTION MASS SPECTROMETRY (HRMS)

The mass spectrometric analysis can be carried out in two different resolutions, namely low resolution and high-resolution mass spectrometry

(HRMS). The LRMS technique gives information about the nominal mass of the analyte, which involves the measurement of m/z values to a single unit (Dass et al., 2007). But the exact mass measurement can only be done using HRMS only, where the mass data acquired with four to six decimal digits (Ekman et al., 2009). The early 1960 have witnessed the evolution of high-resolution mass spectrometers in the analytical field. The critical parameters that define the performance of a mass analyzer include mass range, speed, efficiency, linear dynamic range, sensitivity, resolving power, and mass accuracy (MA) (McLuckey et al., 2001). The basic instrumentation techniques involved in the mass spectral analysis are depicted in Figure 1.1.

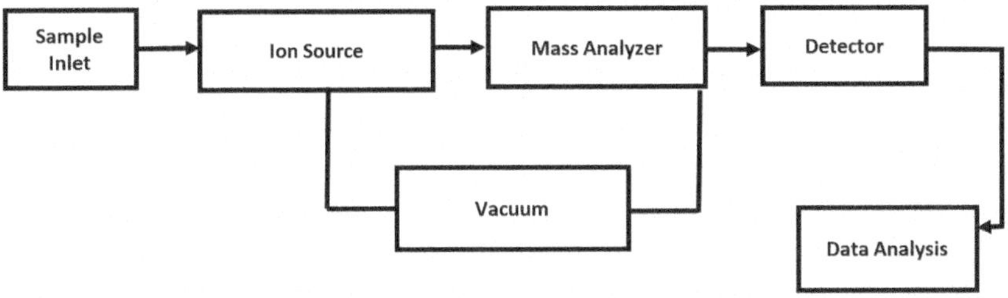

FIGURE 1.1 Basic instrumentation techniques in mass spectrometry.

1.2.2 SAMPLE INTRODUCTION

One of the major struggles in the mass spectral analysis is the sample introduction. For the proper insertion of a sample in mass spectra, which is initially at a high atmospheric pressure (AP), it has to be introduced in such a way that the vacuum inside the chamber remains steadily as required for the instrumental technique. There are mainly two techniques involved, which are direct insertion and direct infusion. Direct insertion techniques involve the introduction of the analyte using an insertion probe followed by insertion into the instrument through a vacuum interlock. The technique involves heating of the sample to have thermal desorption, which is subjected to the high intensity desorption process. Another method is the direct infusion of the analyte of interest through a capillary column or a simple capillary. The great advantage of the direct infusion techniques involves even a small quantity of the sample can be introduced into the mass spectra. Capillary columns are used in the case of gas chromatographic high-resolution mass spectral analysis for the sample introduction techniques. For the sample introduction, there are more methods available. Direct placement of sample (as in DESI), sample with suitable matrix (as in MALDI), direct analysis in real-time (DART), etc., are also employed in various MS techniques.

1.2.3 SAMPLE IONIZATION

To get a better and precise result in MS, the sample ionization technique plays a dominant role. The sample needs to be properly ionized before reaching the detector. Depending upon the mode and ionization, the different ionization techniques are classified as in subsections.

1.2.3.1 THE ELECTROSPRAY IONIZATION PROCESS (ESI)

The basic technique involved in the ESI is the application of electrical energy to assist the movement of ions from analyte solution into the gaseous phase prior to MS analysis. The transfer of ions in the ESI technique consists of multiple steps like dispersal of spray of charged droplets, followed by solvent evaporation and ion ejection from the charged droplets from the charged tube which is kept at a high voltage. The nebulizer gas flowing around the eluted sample allows and triggers a higher sample flow rate, which is followed by a flow towards the analyzer region by a pressure gradient. The charged droplets undergo frequent size reduction with the application of an elevated temperature or nitrogen gas by the evaporation of solvent. ESI technique is more suitable in high to moderate polarity as well as varying molecular weight. In ESI technique, the ionic strength depends on the concentration of the analyte solution and injection parameters (Bruins et al., 1998). ESI MS technique can be successfully used for many clinical applications such as identification of inborn errors of metabolism, including the screening of disorders of amino acids, fatty acid metabolism (Chace et al., 2001), galactosemia (Jensen et al., 2001), cholestatic hepatobiliary disease (Mushtaq et al., 1999), peroxisomal disease (Wanders et al., 1999), disorders in purine and pyrimidine metabolism (Ito et al., 2000), identification of hemoglobin variants, etc. (Wild et al., 2001). One of the drawbacks in the ESI MS is the suppression of the ions in the sample solution because of the matrix interference (Matuszewski et al., 1998).

1.2.3.2 CHEMICAL IONIZATION

Chemical ionization is a technique for forming ions of the analyte of interest by ion/molecule reactions from reactant ions of a reagent gas that is generally present in a much greater abundance than the analyte. Munson and Frank in 1966 introduced the first chemical ionization technique in the field of MS (Alex et al., 2018). In the atmospheric pressure chemical ionization (APCI) technique, comparatively lower energy is required compared to electron ionization (EI). In APCI ionization technique, a flow rate of 200–2,000 μL/min is introduced via a pneumatic nebulizer, which sprays the solution under atmospheric pressure (AP), which then pass through the vaporization chamber to evaporate the solvent. After evaporation of the solvent, the molecules pass through a corona discharge followed by ionization (Horning et al., 1973). The vaporized solvent is ionized to form the reactant ions by the corona discharge followed by ionization of the analyte by charge exchange or proton transfer (Hoffmann et al., 2007). APCI technique is mainly applied to compounds having low to medium polarity and medium molecular weight up to 1,500 Da. The APCI technique has a high degree of tolerability to salts and buffers when compared to ESI techniques (Westman-Brinkmalm et al., 2008). The matrix interference of APCI also showed a high degree of tolerance when compared to ESI (Souverain et al., 2004). The ion suppression factor is also negligible and probably not applicable in APCI technique (Carroll et al., 1975).

1.2.3.3 ATMOSPHERIC PRESSURE PHOTO IONIZATION (APPI)

In APPI technique, the corona discharge in APCI is replaced by the photon-emitting lamp. The technique involves the vaporization of the sample by a heated nebulizer thereby resulting in the formation of gaseous phase molecules of the analyte and the solvent (Raffaelli et al., 2003). Normally krypton lamp is used in APPI techniques. For selective analysis in APPI, the emitted photon energy has to be higher than the ionization energy. This selectivity of APPI technique makes it less susceptible to matrix and salt interferences compared to ESI and APCI (Hanold et al., 2004). The ionization technique in APPI involves two steps which involve direct absorption of the photo energy by the analyte to form the radical cation, which then reacts with the solvent molecule to form the positive hydrogen ion formation. However, direct photoionization happens only in analytes of lower ionization energy than the emitted photon (Syage et al., 2004). Indirect ionization takes place in analytes of higher ionization energy (Marotta et al., 2003).

1.2.3.4 MATRIX-ASSISTED LASER DESORPTION IONIZATION (MALDI)

The basic principle involved in MALDI is photoionization. MALDI technique is based on desorption and ionization of analyte samples incorporated into matrix crystals using a pulsed laser beam (Karas et al., 1987). The critical parameters that influence the MALDI analysis involves the type of matrix, laser parameters and technique of drying (Batoy et al., 2008). In MALDI, the crystals are obtained by spotting a mixture of the sample, which then undergoes air or vacuum drying. The two different mechanisms involved in the MALDI technique are (i) Gas-phase protonation: the first phase involves photoionization of the matrix molecules within the matrix-analyte crystals by the laser, followed by charge transfer from matrix ions to the neutral analyte (Knochenmuss et al., 2006) (ii) The second possible method of ionization involved Lucky Survivor Model. In the Lucky survivor model, the analyte is ionized during sample preparation and incorporated within the sample crystals (Jaskolla et al., 2011). In the refined lucky survivor model, the matrix molecules are incorporated in the desorbed clusters as ionized entities, there by leading to the formation of singly charged analyte ions.

1.2.3.5 AMBIENT IONIZATION METHODS

Compared to the conventional methods of ionization, the ambient ionization methods involve the requirement for sample pretreatment like co-crystallization, including a suitable matrix or dissolving in a suitable solvent. There are now almost 30 different ambient ionization techniques employed till date, which can be classified according to their mode of operation. The most commonly employed ambient methods are discussed in subsections.

1.2.3.5.1 Desorption Electrospray Ionization (DESI)

In the DESI analysis, it is often characterized by the ability to ionize small to large molecules without significant fragmentation (Harris et al., 2008). In the DESI technique electrically charged mixture of solvent and water is subjected to high velocity spraying towards the surface. For high molecular weight samples, the ionization occurs by "droplet pickup" mechanism in which the solvent molecules travel at a higher velocity, thereby occupying a larger diameter, which then undergoes multiple steps to reach the detector (Costa et al., 2008). Generally, DESI method is considered as a wet method as the examined surface is covered with a thin solvent film (Gross, 2014). DESI method is applied in analytical chemistry for the detection of compounds with varying structures (Zhang et al., 2010; Huang et al., 2007; Nyadong et al., 2008).

1.2.3.5.2 Direct Analysis in Real-Time (DART)

Cody et al. introduced the DART technique in 2005 as a solvent free method and thus there is no need for solvent for the ionization (Cody et al., 2005). The method usually depends on the collision of the examined surface using a hot stream of gas, which is flowing at a rate of 1 L/min which carries the active components. The source tube consists of several chambers with gases, nitrogen, and helium. The electrical voltage applied in the preliminary chamber comprising of the needle and perforated disc subjected to the electrical glow, followed by the production of ions, electrons, and excited ions (Alberici et al., 2010). The use of DART for quantification in liquid samples is more accurate than solid matrices in spite of the irreproducibility in liquid sample positioning.

1.2.4 RESOLUTION POWER, RESOLUTION, AND MASS ACCURACY (MA)

Resolution of an instrument refers to the capability of the instrument to properly discriminate ions of similar m/z ratio. The mass resolution is generally defined as $m/\Delta m$; where 'm' is the mass of the ion and 'Δm' is either the peak width or the spacing between two either equal peaks or is generally termed as 10% valley definition. The above-mentioned description is further modified as full width at half maximum (FWHM), in which 'm' is defined as mass of the ion corresponding to the peak and 'Δm' as the full width of the peak in the mass units corresponding to the half the maximum height of the peak. The FWHM has the advantage, as it is largely independent of peak shape and can be identified even if the baseline is noisy.

MA of a technique refers to the accuracy with which the m/z value of the peak can be determined. The MA is a characteristic of the mass analyzer and is dominated by scan-to-scan variation in the m/z values. The difference between the actual value and experimentally calculated value is normally expressed in terms of milli Dalton (mDa) or parts per million (ppm). To determine the MA, exact mass and mass defect needs to be defined and determined. Exact mass is determined as the mass of the ion with known charge (Urban et al., 2014) and mass defect is defined as the calculated mass of an ion/molecule with specific isotopic composition (Zhu et al., 2006).

1.2.5 MASS CALIBRATION

For every sensitive instrument, mass calibration plays a dominant role and is done using mass reference compounds with known m/z values (Busch et al., 2002). The calibration is performed in either an automatic or a semi-automatic manner and is generally referred to as external calibration. The critical parameters involved in the selection of the mass calibration compounds include (i) it should regularly yield abundant ions across the entire scan range, (ii) the reference compounds should possess negative mass defects to adjust the overlap with atoms like C, H, O, and N (iii) has to be stable and volatile (Dass et al., 2007).

1.2.6 TARGETED AND NON-TARGETED DATA ACQUISITION

In the targeted mode of data acquisition, the analyzer is programmed in such a way as to identify only pre-determined molecules of interest. The targeted mode of data acquisition is usually performed using triple quadrupole mass spectrometers. In the non-targeted data acquisition method, the critical parameters involved include peak extraction, peak annotation and alignment, various statistical analysis, etc. Full-scan, data-dependent acquisition (DDA), and data-independent acquisition (DIA) are the common methods involved in data acquisition in non-targeted analysis in MS. Normally the DDA method is involved for a time efficient approach for the data acquisition. In the DDA analysis mode, the MS Spectrum undergoes full scan analysis followed by MS analysis with precursor ions obtained from the full spectrum, which allows both quantitative and structural information about the molecule of interest (Benton et al., 2015). But the DDA method lacks the efficiency of detection of low concentrations of metabolites. To overcome the limitations of the DDA, DIA method was introduced to elucidate more information on the metabolic profile even at lower concentrations (Schrimpe et al., 2016). The most common method employed include all ion fragmentation, in which all precursor ions are fragmented, followed by a sequential window acquisition of all the theoretical fragment-ion spectra (SWATH) (Roemmelt et al., 2014). Several methods have been developed to address this goal. The most common methods include XCMS online, MZmine, MetAlign, etc. The XCMS online service includes compiling of the raw data processing, retention time (RT) correction calculations and metabolite assignment (Tauten-Hahn et al., 2012). The MZmine software is designed in such a way to efficiently separate core functionality and data processing modules. In MZmine, data processing modules use embedded visualization tools to have a quick view on the processing parameters (Pluskal et al., 2010).

1.2.7 COMPOUND IDENTIFICATION

The compound identification in mass spectra includes comparing the spectra generated from unknown samples in reference with the data base containing spectra of compounds in the library (Colby et al., 2017). The basic acquisition methods in mass spectra include full-spectrum mode and selected ion mode. The first strategy developed for the data acquisition in MS is DDA, which involves the selection of the precursor ions for fragmentation relative to their abundances. Another method involved is data-independent acquisition (DIA) which involves scanning of the precursor ions with a m/z window regardless of their intensity. Dong et al. (2007) studied that DIA produced better quantitative results with good signal to noise ratio (S/N), which provides a better quantification of peptides and proteins with better reproducibility. Another method involved is the Sequential window acquisition of all theoretical fragment ion spectra (SWATH) which involves data acquisition within a small range followed by fragmentation and without any loss of spectral data which then undergoes further fragmentation and mass range scanning until the desired range is achieved (Zhu et al., 2014). Collision-induced dissociation (CID) of an isolated ion species in mass spectra is a powerful technique for the determination of both ion structures and the analytical identification of compounds with high specificity (Haag, 2016). The molecular ions and the fragmented ions are analyzed using either an inbuilt library or a primary reference standard.

1.3 CONCLUSION

Mass spectral analysis is regarded as one of the sophisticated instruments, which is widely used in a wide variety of applications. The high degree of sensitivity of the HRMS-QTOF provides a valuable tool for the identification of trace elements and molecules whose presence cannot be determined using other mass spectral instruments. HRMS is an important tool in the research field as it helps to identify the unknown molecules. The use of HRMS can be widely explored in various fields like clinical and forensic toxicology, drug metabolite identification, identification of unknown substances, etc.

KEYWORDS

- **data acquisition**
- **Fourier transform orbitrap**
- **high-resolution mass spectrometry**
- **ionization techniques**
- **mass accuracy**
- **resolution**
- **sector mass spectrometer**

REFERENCES

Alberici, R. M., Simas, R. C., Sanvido, G. B., Romao, W., Lalli, P. M., Benassi, M., Cunha, I. B., & Eberlin, M. N., (2010). Ambient mass spectrometry: Bringing MS into the "real world." *Anal. Bioanal. Chem., 398*, 265–294.

Alex, G. H., (2018). *Chemical Ionization Mass Spectrometry.* CRC Press: Routledge, United Kingdom.

Batoy, S. M. A. B., Akhmetova, E., Miladinovic, S., Smeal, J., & Wilkins, C. L., (2008). Developments in MALDI mass spectrometry: The quest for the perfect matrix. *Appl. Spectros. Rev., 43*, 485–550.

Benton, H. P., Ivanisevic, J., Mahieu, N. G., Kurczy, M. E., Johnson, C. H., Franco, L., et al., (2015). Autonomous metabolomics for rapid metabolite identification in global profiling. *Anal Chem., 87*, 884–891.

Bruins, A. P., (1998). Mechanistic aspects of electrospray ionization. *J. Chromatogr A., 794*, 345–357.

Busch, K. L., (2002). SAMS: Speaking with acronyms in mass spectrometry. *Spectroscopy, 17*, 55–62.

Busch, K. L., Glish, G. L., & McLuckey, S. A., (1998). *Mass Spectrometry/Mass Spectrometry: Techniques and Applications of Tandem Mass Spectrometry*. VCH, New York.

Carroll, D., Dzidic, I., Stillwell, R., Haegele, K., & Horning, E., (1975). Atmospheric pressure ionization mass spectrometry. Corona discharge ion source for use in a liquid chromatograph-mass spectrometer–computer analytical system. *Anal. Chem., 47*, 2369–2373.

Chace, D. H., DiPerna, J. C., Mitchell, B. L., Sgroi, B., Hofman, L. F., & Naylor, E. W., (2001). Electrospray tandem mass spectrometry for analysis of acylcarnitines in dried postmortem blood specimens collected at autopsy from infants with unexplained cause of death. *Clin Chem., 47*, 1166–1182.

Cody, R. B., Laram´ee, J. A., & Durst, H. D., (2005). Versatile new ion source for the analysis of materials in open air under ambient conditions. *Anal. Chem., 77*, 2297–2302.

Colby, J. M., Rivera, J., Burton, L., Cox, D., & Lynch, K. L., (2017). Improvement of drug identification in urine by LC-QqTOF using a probability-based library search algorithm. *Clin. Mass Spectrom., 3*, 7–12.

Costa, A. B., & Cooks, R. G., (2008). Simulated splashes: Elucidating the mechanism of desorption electrospray ionization mass spectrometry. *Chem. Phys. Lett., 464*, 1–8.

Dass, C., (2007). *Fundamentals of Contemporary Mass Spectrometry.* John Wiley & Sons: Hoboken, New Jersey.

Dong, M. Q., Venable, J. D., Au, N., Xu, T., Park, S. K., Cociorva, D., Johnson, J. R., et al., (2007). Quantitative mass spectrometry identifies insulin signaling targets in *C. elegans*. *Science, 317*, 660–663.

Ekman, R., Silberring, J., Westman-brinkmalm, A., & Kraj, A., (2009). *Mass Spectrometry: Instrumentation, Interpretation, and Applications* (p. 372). John Wiley & Sons.

Gross, J. H., (2014). Direct analysis in real-time—A critical review on DART-MS. *Anal. Bioanal. Chem., 406*, 63–80.

Haag, A. M., (2016). Mass analyzers and mass spectrometers. *Adv. Exp. Med. Biol., 919*, 157–169.

Hanold, K. A., Fischer, S. M., Cormia, P. H., Miller, C. E., & Syage, J. A., (2004). Atmospheric pressure photoionization. 1. General properties for LC/MS. *Anal. Chem., 76*, 2842–2851.

Harris, G. A., Nyadong, L., & Fernandez, F. M., (2008). Recent developments in ambient ionization techniques for analytical mass spectrometry. *Analyst, 133*, 1297–1301.

Hoffmann, E., & Stroobant, V., (2007). *Mass Spectrometry: Principles and Applications* (3rd edn.). John Wiley: Chichester, West Sussex, UK.

Horning, E., Horning, M., Carroll, D., Dzidic, I., & Stillwell, R., (1973). New pictogram detection system based on a mass spectrometer with an external ionization source at atmospheric pressure. *Anal. Chem., 45*, 936–943.

Hu, Q., Noll, R. J., Li, H., Makarov, A., Hardman, M., & Graham, C. R., (2005). The orbitrap: A new mass spectrometer. *J. Mass Spectrom., 40*, 430–443.

Huang, G., Chen, H., Zhang, X., Cooks, R. G., & Ouyang, Z., (2007). Rapid screening of anabolic steroids in urine by reactive desorption electrospray ionization. *Anal. Chem., 79*, 8327–8332.

Ito, T., Van, K. A. B. P., Bootsma, A. H., Haasnoot, A. J., Cruchten, A. V., Wada, Y., & Van, G. A. H., (2000). Rapid screening of high-risk patients for disorders of purine and pyrimidine metabolism using HPLC-electrospray tandem mass spectrometry of liquid urine or urine-soaked filter paper strips. *Clin Chem., 46*, 445–452.

Jaskolla, T. W., & Karas, M., (2011). Compelling evidence for lucky survivor and gas-phase protonation: The unified MALDI analyte protonation mechanism. *J. Am. Soc. Mass. Spectrom., 22*, 976–988.

Jensen, U. G., Brandt, N. J., Christensen, E., Skovby, F., Nogaard-Pedersen, B., & Simonsen, H., (2001). Neonatal screening for galactosemia by quantitative analysis of hexose monophosphate using tandem mass spectrometry: A retrospective study. *Clin. Chem., 47*, 1364–1372.

Karas, M., Bachmann, D., Bahr, U., & Hillenkamp, F., (1987). Matrix-assisted ultraviolet laser desorption of non-volatile compounds. *Int. J. Mass Spectrom. Ion. Process., 78*, 53–68.

Katajamaa, M., & Orešič, M., (2005). Processing methods for differential analysis of LC/MS profile data. *BMC Bioinformatics, 6*, 179.

Knochenmuss, R., (2006). Ion formation mechanisms in UV-MALDI. *Analyst, 131*, 966–986.

Marotta, E., Seraglia, R., Fabris, F., & Traldi, P., (2003). Atmospheric pressure photoionization mechanisms: 1. The case of acetonitrile. *Int. J. Mass Spectrom., 228*, 841–849.

Matuszewski, B. K., Constanzeer, M. L., & Chavez-Eng, C. M., (1998). Matrix effect in quantitative LC/MS/MS analyses of biological fluids: A method for determination of finasteride in human plasma at picogram per milliliter concentrations. *Anal. Chem., 70*, 882–889.

McLuckey, S. A., & Wells, J. M., (2001). Mass analysis at the advent of the 21st century. *Chem. Rev., 101*, 571–560.

Mushtaq, I., Logan, S., Morris, M., Johnson, A. W., Wade, A. M., Kelly, D., & Clayton, P. T., (1999). Screening of newborn infants for cholestatic hepatobiliary disease with tandem mass spectrometry. *BMJ, 319*, 471–477.

Nyadong, L., Hohenstein, E. G., Johnson, K., Sherrill, C. D., Green, M. D., & Fern´andez, F. M., (2008). Desorption electrospray ionization reactions between host crown ethers and the influenza neuraminidase inhibitor oseltamivir for the rapid screening of Tamiflu. *Analyst, 133*, 1513–1522.

Petras, D., Jarmusch, A. K., & Dorrestein, P. C., (2017). From single cells to our planet–recent advances in using mass spectrometry for spatially resolved metabolomics, *Curr. Opin. Chem. Biol., 36,* 24–31.

Pluskal, T., Castillo, S., Villar-Briones, A., & Oresic, M., (2010). MZmine 2: Modular framework for processing, visualizing, and analyzing mass spectrometry-based molecular profile data. *BMC Bioinformatics, 11*, 395.

Raffaelli, A., & Saba, A., (2003). Atmospheric pressure photoionization mass spectrometry. *Mass Spectrom. Rev., 22*, 318–331.

Roemmelt, A. T., Steuer, A. E., Poetzsch, M., & Kraemer, T., (2014). Liquid chromatography, in combination with a quadrupole time-of-flight instrument (LC QTOF), with sequential window acquisition of all theoretical fragment-ion spectra (SWATH) acquisition: Systematic studies on its use for screenings in clinical and forensic toxicology and comparison with information-dependent acquisition (IDA). *Anal. Chem., 86*, 11742–11749.

Schrimpe-Rutledge, A. C., Codreanu, S. G., Sherrod, S. D., & McLean, J. A., (2016). Untargeted metabolomics strategies—challenges and emerging directions. *J. Am. Soc. Mass Spectrom., 27*, 1897–1905.

Souverain, S., Rudaz, S., & Veuthey, J. L., (2004). Matrix effect in LC-ESI-MS and LC-APCIMS with off-line and on-line extraction procedures. *J. Chrom., 1058*, 61–66.

Syage, J. A., (2004). Mechanism of $[M+ H]^+$ formation in photoionization mass spectrometry. *J. Am. Soc. Mass Spectrom., 15*, 1521–1533.

Tauten-Hahn, R., Patti, G. J., Rinehart, D., & Siuzdak, G., (2012). XCMS online: A web-based platform to process untargeted metabolomic data. *Anal Chem., 84*, 5035–5039.

Urban, P. L., (2016). Quantitative mass spectrometry: An overview. *Philos. Trans. A Math. Phys Eng. Sci., 374*, 20150382.

Wanders, R. J. A., Schutgens, R. B., & Barth, P. G., (1995). Peroxisomal disorders: A review. *J. Neuropathol. Exp. Neurol., 54*, 726–739.

Westman-Brinkmalm, A., & Brinkmalm, G., (2008). A mass spectrometer's building blocks. In: Ekman, R., Silberring, J., Westman-Brinkmalm, A., & Kraj, A., (eds.), *Mass Spectrometry* (pp 24, 25). Wiley: Hoboken, NJ.

Wild, B. J., Green, B. N., Cooper, E. K., Lalloz, M. R., Erten, S., Stephens, A. D., & Layton, D. M., (2001). Rapid identification of hemoglobin variants by electrospray ionization mass spectrometry. *Blood Cells Mol. Dis., 27*, 691–704.

Zhang, Y., & Chen, H., (2010). Detection of saccharides by reactive desorption electrospray ionization (DESI) using modified phenylboronic acids. *Int. J. Mass Spectrom., 289*, 98–107.

Zhu, M., Ma, L., Zhang, D., Ray, K., Zhao, W., Humphreys, W. G., Skiles, G., Sanders, M., & Zhang, H., (2006). Detection and characterization of metabolites in biological matrices using mass defect filtering of liquid chromatography/high resolution mass spectrometry data. *Drug Metab. Dispos., 34*, 1722–1733.

Zhu, X., Chen, Y., & Subramanian, R., (2014). Comparison of information-dependent acquisition, SWATH, and MSAll techniques in metabolite identification study employing ultrahigh-performance liquid chromatography–quadrupole time-of-flight mass spectrometry. *Anal. Chem., 86*, 1202–1209.

CHAPTER 2

Identification and Structure Elucidation of Compounds: Deleting 'Un' from the Unknowns

JOBY JACOB, SHINTU JUDE, and SREERAJ GOPI

R&D Center, Aurea Biolabs (P) Ltd., Kolenchery, Cochin – 682311, Ernakulam, Kerala, India, E-mail: jobyjacob799@gmail.com (J. Jacob)

ABSTRACT

The occurrence of unknown compounds is a major consideration in many analytical fields. HRMS forms a basis of unknown identification analysis procedures, in accordance with the peculiar properties such as exact mass determination and high resolution (HR). Moreover, the developments in the field of data processing software have reduced the complications in the domain. The chapter elicits the general principles of unknown identification analysis and discusses about some major sectors, where unknown identification is an inevitable part. Some recent examples are selected so as to deliver a clear idea on different approaches and strategies adopted in the field.

2.1 INTRODUCTION

Unknown identification can be of two types: known unknowns (also termed as unknown knowns) or unknown unknowns. Known unknowns

High-Resolution Mass Spectrometry and Its Diverse Applications: Cutting-Edge Techniques and Instrumentation. Sreeraj Gopi, PhD, Sabu Thomas, PhD, Augustine Amalraj, PhD, & Shintu Jude, MSc (Editors)

may not be entirely new compounds, though unknown to the analyst can be identified by non-target analysis. Unknown-unknown compounds are actually the new compounds, identification of which includes structure elucidation also (Little et al., 2011). The basic idea to figure out an unknown compound is its mass. So, exact mass measurement along with high resolution (HR) is a perfect solution for unknown identification. Hence, nowadays, high-resolution mass spectrometry (HRMS) is considered as the major technique to characterize unknown compounds. The possibilities of HRMS analysis modes can be targeted and untargeted, routine and research, quantitative or qualitative analyzes and are able to detect the presence, level, or fate of compounds of interest (Li et al., 2016). They alone can provide a huge amount of information and are helpful in deriving the empirical formula. In the case of volatile compounds, the reference libraries of EI mass spectra provide adequate knowledge on the known unknowns. But for non-volatile compounds, modifications such as hyphenation with chromatography and libraries of product mass spectra are recommended (Milman, 2015). Along with other techniques, majorly nuclear magnetic resonance (NMR), HRMS can synergistically present much more structural information. Liquid chromatography (LC)-HRMS-NMR is a well-established technique for the small molecule identification.

HRMS offers a platform for the non-targeted screening of unknown sample matrices, which facilitates the addressing of compounds without any prior information. For non-targeted screening, three different approaches: matching theoretical reference data, comparison to the predicted reference data and data interpretation- are usually accepted (Milman, 2011). HRMS acquisition is always associated with a high amount of data, and powerful data processing tools are required for the proper utilization of the data generated. Development of powerful data processing software and systems pave the way for new strategies and approaches in this field (Agüera et al., 2017).

The strategy involves the detection of a maximum number of peaks from the raw chromatogram and gathering information such as retention time (RT), m/z values, signal intensity, etc., from the same. Highly resolved peaks with accurate mass measurement allow the detection of possible molecular formulas and isotope patterns. MS/MS fragmentations are possible, which in turn serve confirmation purposes. Together, all these data can be utilized for the identification of candidate structures, by utilizing established databases and libraries. There are databases available such as

Chemspider to facilitate this step. Normally, possibilities of hundreds to thousands of candidates may arise (Little et al., 2012; Hu et al., 2018).

The second stage in the unknown screening is to prioritize the candidate structures. This can be proceeded by comparing their structure-based predictions and their analytical information. Usually, the in silico fragmentation pattern is compared with the experimental MS/MS spectra and a similarity index is prepared out of these results. The proposed candidate structures are ranked, according to the similarity between the experimental and theoretical spectra (Hu et al., 2018). In order to help the scenario, there are many prediction systems and approaches developed and are classified as rule-based (e.g., Mass Frontier, ACD/MS Fragmenter, MOLGEN-MS, etc.), combinational (e.g., EPIC, Metfrag, MetFusion, etc.), machine learning, etc. (Hu et al., 2018; Vaniya and Fiehn, 2015; Ruttkies et al., 2016).

RT based prediction systems are also present in the picture. Chromatographic RT is an experimental information, and is available from the acquired data. Based on the quantitative structure-retention relationships, different RT prediction models have been developed. These models are executed by comparing the predicted RT, to the experimental RT. Generally, the RT prediction systems work well for neutral and less hydrophilic compounds, than ionic or strong hydrophilic compounds. The special features in the new generation HRMS instrumentation allow the accurate measurement of relative isotope abundances for small molecules. These algorithms are considered isotope ratios for the elemental compositions, providing one more confirmation step for the proposed compound (Tyrkkö et al., 2012; Bade et al., 2015).

The major hurdle in unknown analysis is to evaluate the probability of misidentification. There are various parameters influencing the occurrence of false positive, such as compound composition, presence of isomers, molecular mass, mass defect, matrix effects, mode of analysis, chosen databases and properties of detection specificity including mass accuracy (MA) and mass resolution (Rochat, 2017). Of course, chemical composition or accurate m/z determination is a foundation for unknown identification. But, the usage of relative RT, ion ratios, fragmentation of molecular ions, etc., increases the possibilities for a robust and confident identification (Kind and Fiehn, 2006; Weber and Viant, 2010). A combination of different prediction models and their comparison with corresponding experimental data act complementary and improves the ranking of correct candidate, which in turn results in the proper identification of unknown compounds.

The HR analytical instruments allow HR – full scan runs, which facilitates the redefining of pre-acquisition choices by permitting the retrospective analysis of acquired data (Rochat, 2017). Beneficial role of unknown identification is worth in many analytical tactics. The untargeted analyzes allow the identification of both biomarkers and toxic/illicit compounds. This is important in food safety, environmental, clinical, forensic, security, antidoping, and toxicology analyzes (De Cássia Lemos Lima et al., 2018; Liu et al., 2019; Prakash et al., 2019; Shaikh et al., 2020). Some of the important fields, where the unknown identification functions as the cornerstone are discussed in the chapter.

2.2 PLANT-BASED MOLECULES

Molecules which are extracted from plants and herbs are considered in the prevention of many ailments such as cancer, heart disease, and other diseases with health-promoting, disease-preventing, or curative properties. These molecules have enormous properties such as antioxidant, anti-inflammatory, immunity enhancement and other potential health benefits. In this context, the well thorough knowledge of these molecules are very important to understand the biological pathways and the mode of action in the body. At present, researchers have identified hundreds of health promoting molecules, and new discoveries concerning the complex interactions between secondary metabolites in plants and health are continually made. The search of plant molecules with medicinal properties is still ongoing and some of the important and recent discoveries are depicted below.

Tuenter et al. isolated seven cyclopeptide alkaloids from the stem bark of *Ziziphus nummularia* and *Ziziphus spina-christi*. The identification of molecules were done by HPLC-PDA-(HRMS)-SPE-NMR after the purification by semi-preparative HPLC. From the purified extract, nummularine-U, spinanine-B, and spinanine-C were identified as unknown and mauritine-F, nummularine-D, nummularine-E, and amphibine-D as known compounds. The elucidated structures are presented in Figure 2.1 (Tuenter et al., 2017).

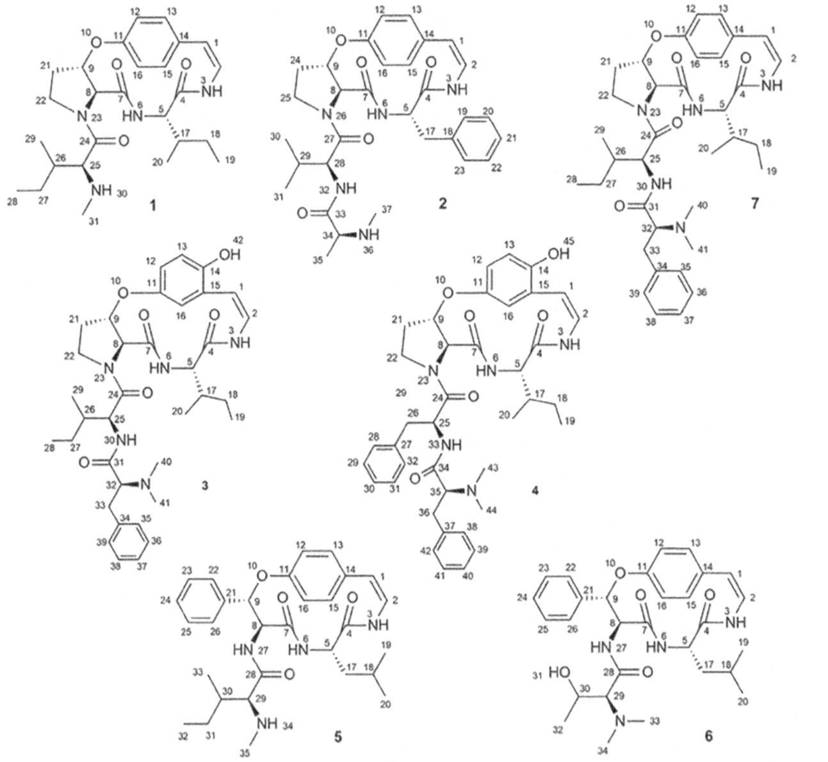

FIGURE 2.1 Chemical structures of compounds form *Z. spina-christi* and *Z. nummularia*: (1)nummularine-U; (2) mauritine-F; (3) spinanine-B; (4) spinanine-C; (5) nummularine-D; (6) nummularine-E; and (7) amphibine-D.

Source: Reprinted with permission from Tuenter et al., 2017. © Elsevier.

Different types of lipids constitute a key role in coffee bean process, coffee brew and in the effects of coffee on human health. Carolina et al. established a protocol to identify the lipid profile using LC-HRMS/MS in green Arabica coffee beans. They have used three different methods to extract the samples for analysis, namely, Bligh-dyer (BD), Folch (FO), and Matyash (Ma). In which, the Ma method gives the maximum number of lipids (131 lipids) in comparison with other methods (Silva et al., 2020). In this method, the immature beans showed lower levels of phosphatidylinositol, phosphatidylethanolamine, phosphatidylcholine, lysophosphatidylcholine, $^{\beta}$N-alkanoyl-5-hydroxytryptamides and lysophosphatidylethanolamine compared to mature beans (Figures 2.2 and 2.3) (Silva et al., 2020).

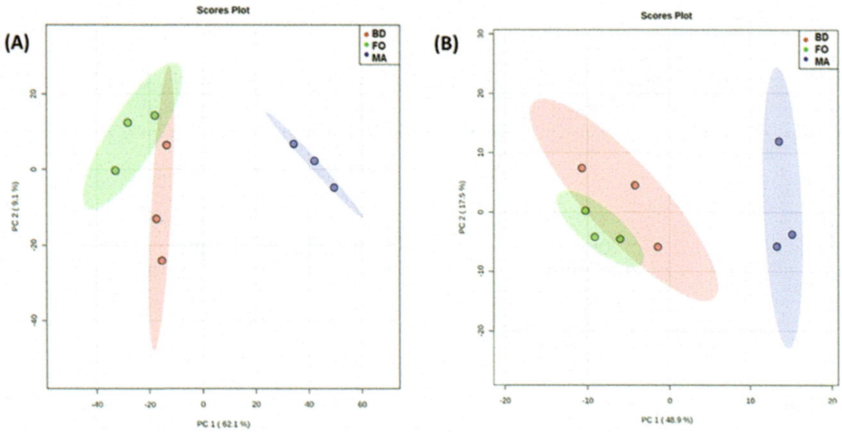

FIGURE 2.2 Principal component analysis from LC-HRMS data of lipid extractions. (A) data from the positive; and (B) negative ESI modes. Red spots: Bligh-dyer (BD); Green spots: Folch (FO), Blue spots: Matyash (Ma) extraction methods.

Source: Reprinted with permission from Silva et al., 2020. © Elsevier.

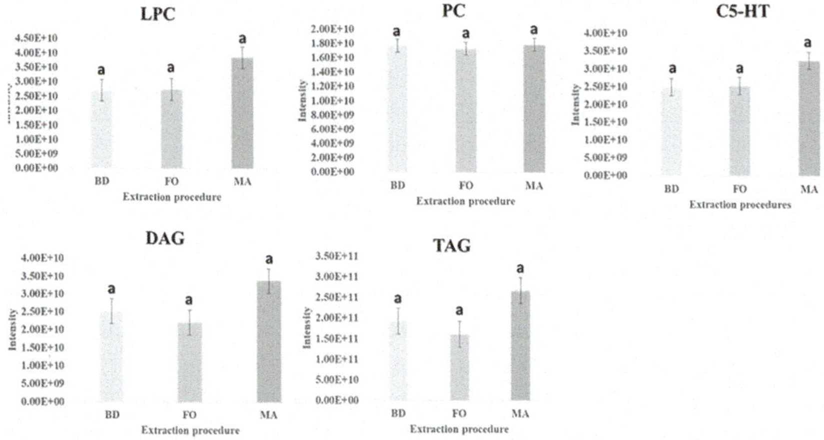

FIGURE 2.3 Comparison of the sum of peak areas from lipids of different classes detected in positive ESI mode. [LPC: lysophosphatidylcholine; PC: phosphatidylcholine; DAG: diacylglycerol; TAG: triacylglycerol; C-5HT: βN-alkanoyl-hydroxytryptamide; BD: Bligh-dyer; FO: Folch; Ma: Matyash. Small letters indicate significance at $p < 0.05$].

Source: Reprinted with permission from Silva et al., 2020. © Elsevier.

The adulteration of herbal formulations also can be found using LC-HRMS and LC-MS-SPE/NMR. For example, Kesting et al. analyzed the Gold Nine Soft Capsules, a Chinese herbal-based medicine proposed to treatment of hypertension, to identify the known anti-hypertensive drugs, namely, amlodipine, valsartan, and indapamide, which were not declared on the label. They found these adulterants in the herbal formulation by LC-HRMS in this case study (Kesting et al., 2010). The comparative chromatograms of Gold Nine Soft Capsules and standards are given in Figure 2.4.

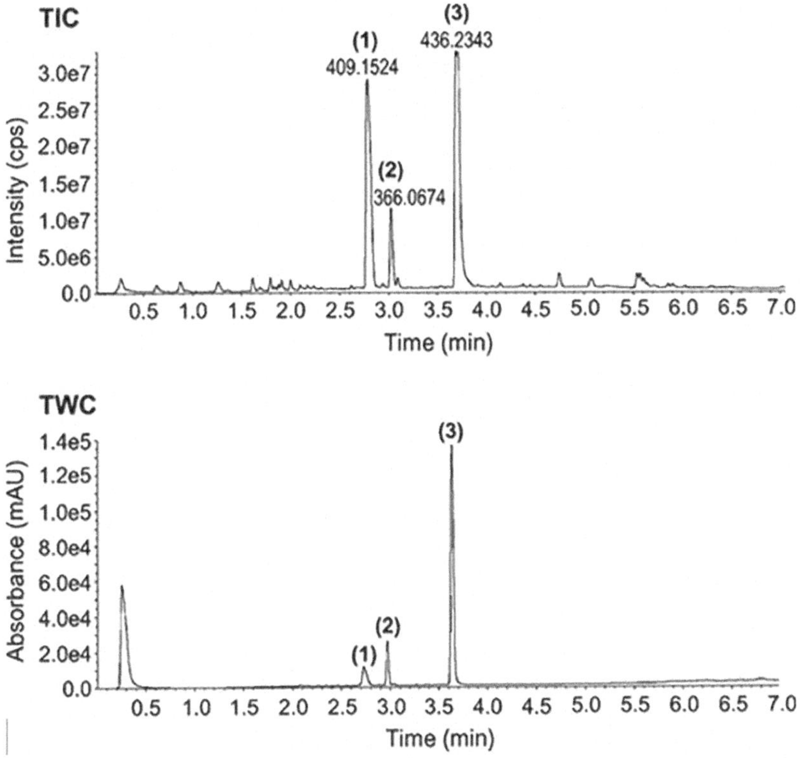

FIGURE 2.4 LC-HRMS of gold nine soft capsules. Top: total ion chromatogram, positive mode, with high-resolution mass indicated for each peak. Lower: total wavelength chromatogram (UV). (1) Amlodipine; (2) indapamide; and (3) valsartan.

Source: Reprinted with permission from Silva et al., 2020. © Elsevier.

In the wine industry, sulfur chemistry is an important factor to determine the quality of wine at any stage of the production process. Hence, van Leeuwen et al. developed a method by UHPLC-HRMS to analyze glutathionyl and cysteinyl polysulfides in wine. These polysulfides contains 3 to 5 sulfur atoms, and they have compared with synthesized reference standards and identified 11 different glutathionyl and cysteinyl polysulfides in different wines (van Leeuwen et al., 2020). The score parameters for the identification using Compound Discoverer™ 3.1 is given in Table 2.1.

TABLE 2.1 Score Parameters of Polysulfide Identification Using Compound Discoverer™ 3.1

Compound	Predicted Neutral Chemical Formula	Molecular Weight (Da)	Pattern Coverage (%)	FISh Coverage (%)
CSSC	$C_6H_{12}N_2O_4S_2$	240.02385	99.21	91.11
CSSSC	$C_6H_{12}N_2O_4S_3$	271.99592	100.00	92.43
CSSSSC	$C_6H_{12}N_2O_4S_4$	303.9674	99.85	100.0
CSSSSSC	$C_6H_{12}N_2O_4S_5$	335.9395	98.1	32.43
GSSG	$C_{20}H_{32}N_6O_{12}S_2$	612.15037	99.63	65.35
GSSSG	$C_{20}H_{32}N_6O_{12}S_3$	644.1235	100.00	70.27
GSSSSG	$C_{20}H_{32}N_6O_{12}S_4$	676.0955	99.91	53.33
GSSSSSG	$C_{20}H_{32}N_6O_{12}S_5$	708.0876	100.00	34.21
CSSG	$C_{13}H_{22}N_4O_8S_2$	426.0868	95.00	96.09
CSSSG	$C_{13}H_{22}N_4O_8S_3$	458.0586	100.00	83.64
CSSSSG	$C_{13}H_{22}N_4O_8S_4$	490.0311	100.00	50.00

Note: GS: glutathione; CS: cysteine.

FISh coverage = *($\sum_{per\ all\ scans}$ # matched centroids)* × *100; where*: ($\sum$ per all scans # used centroids)

 # matched centroids constitute the number of matched centroids.

 # used (matched + unmatched) centroids constitutes the number of centroids in the fragmentation scan that are above the operator defined signal-to-noise threshold. The algorithm bypasses centroids lower than the operator-defined signal-to-noise threshold.

Source: Reprinted with permission from van Leeuwen et al., 2020. © Elsevier.

Table 2.2 describes different HRMS configurations used for the identification of the phytochemical composition of different plants and the molecules identified.

TABLE 2.2 The HRMS Methods to Identify the Molecules from Different Plants

Plant Used	Plant Part Used	Method	Molecules Identified	References
Paeonia rockii	Leaves and Flowers	Q-exactive orbitrap hybrid quadrupole-orbitrap mass spectrometry	16 compounds were reported for the first time	Li et al. (2016)
Aechmea magdalenae	Rhizomes	LTQ orbitrap discovery in positive ion mode	Resazurin was reduced to resorufin	Monga et al. (2019)
Papaver nudicaule L.	Aerial parts	liquid chromatography high-resolution mass spectrometry and molecular networking	Isoquinoline alkaloids	Song et al. (2020)
Eremanthus crotonoides	Leaves	HPLC-HRMS-SPE-NMR	Six α-glucosidase inhibitors, namely quercetin, trans-tiliroside, luteolin, quercetin-3-methyl ether, 3,5-di-O-caffeoylquinic acid n-butyl ester and 4,5-di-O-caffeoylquinic acid n-butyl ester	Lobo et al. (2016)
Polygonum cuspidatum	Root	HPLC-HRMS	Procyanidin B2 3,3″-O-digallate (3) and (–)-epicatechin gallate	Zhao et al. (2017)
Artemisia annua	Aerial parts	UHPLC-HRMS	Chrysosplenol D	Messaili et al. (2020)

TABLE 2.2 *(Continued)*

Plant Used	Plant Part Used	Method	Molecules Identified	References
Rorippa indica	–	UHPLC-PDA-ESI/HRMS^n	Glucosinolates, containing 24 new glucosinolates, including 14 glucosylated glucosinolates	Lin et al. (2014)
Hancornia speciosa	Leaves	UHPLC coupled to orbitrap HRMS	17 new molecules, protocatechuic acid, catechin, and quercetin, and another 14 were putatively identified viz. B- and C-type procyanidins	Bastos et al. (2017)
Miconia albicans	Leaves	HPLC-HRMS-SPE-NMR	Protein tyrosine phosphatase 1B inhibitors including 1-O-(E)-caffeoyl-4,6-di-O-galloyl-β-D-glucopyranose, myricetin 3-O-α-L-rhamnopyranoside, quercetin 3-O-(2”-galloyl)-α-L-rhamnopyranoside, mearnsetin 3-O-α-L-rhamnopyranoside, kaempferol 3-O-α-L-arabinopyranoside, maslinic acid, 3-epi-sumaresinolic acid, sumaresinolic acid, 3-O-cis-p-coumaroyl maslinic acid, 3-O-trans-p-coumaroyl maslinic acid, 3-O-trans-p-coumaroyl 2α-hydroxydulcioic acid, oleanolic acid and ursolic acid	De Cássia Lemos Lima et al. (2018)

TABLE 2.2 *(Continued)*

Plant Used	Plant Part Used	Method	Molecules Identified	References
Diaphragma juglandis	Fructus	UHPLC-Q-orbitrap HRMS	Around 200 compounds, including hydrolyzable tannins, flavonoids, phenolic acids, and quinones	Liu et al. (2019)
Pyrus pashia	Fruits	UPLC-ESI-HRMS/MS	Benzoic, cinnamic acid derivatives and flavonoids. Hydroquinone, chlorogenic acid, arbutin, and catechin	Prakash et al. (2019)
Ficus religiosa	Leaves	LC-HRMS	Acipimox, edoxudine, levulinic acid, hydroxyhydroquinone, ramiprilglucuronide, berberine, antimycin A, swietenine, and some short peptides	Shaikh et al. (2020)

2.3 METABOLIC PATHWAYS

Metabolism is defined as a set of life-sustaining chemical reactions in organisms by the chemical and physical process participating in the maintenance of life. It consists of a wide range of enzymatic processes and transport mechanisms used to transform thousands of organic molecules into several different molecules essential to maintain cellular life (Schilling et al., 2000). HRMS can be used to find out various metabolic pathways of drugs, toxic molecules, and phytochemicals in the blood, urine, hair, liver, and to track fluid metabolic profiling. Martín-Blázquez et al. identified a differential metabolomic signature for metastatic colorectal cancer by analyzing the serum samples using LC-HRMS of colorectal cancer patients with control. This study revealed that the LC-HRMS is a potent diagnostic tool for metastatic colorectal cancer (Martín-Blázquez et al., 2019). Similarly, the metabolism of drugs like vardenafil was also analyzed by the *in vitro* transformation with the use of human liver microsomes in comparison to TiO_2 and WO_3 assisted photocatalytic method and the samples were analyzed by LC-ESI-HRMS (Figure 2.5) (Gawlik and Skibiński, 2018).

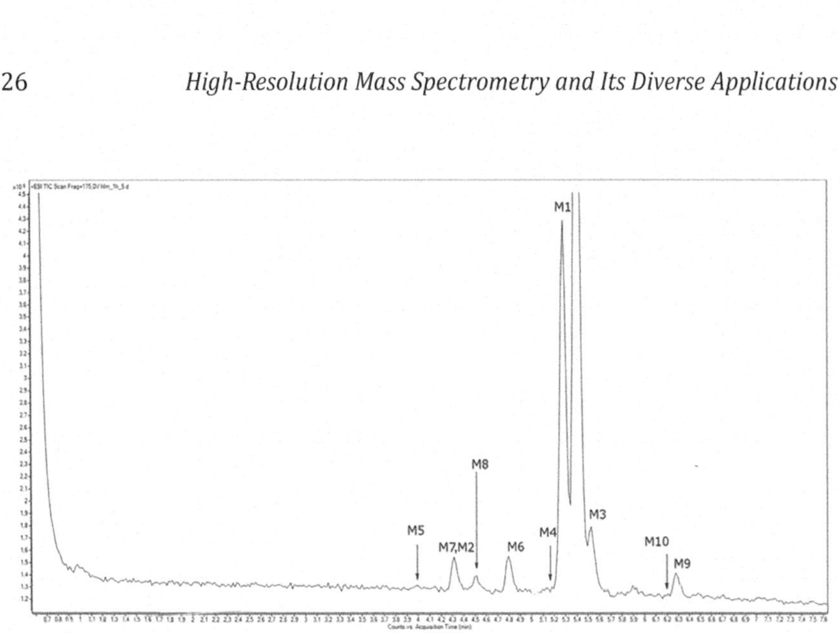

FIGURE 2.5 Example total ion chromatogram (TIC) obtained for vardenafil after 60 min incubation with human liver microsomes.

Source: Reprinted with permission from Gawlik and Skibiński, 2018. © Elsevier

Chekmeneva et al. developed a method to detect over 40 urinary metabolites using direct infusion nanoelectrospray high-resolution mass spectrometry (DI-nESI-HRMS) method with time-of-flight detection for rapid analysis. The advantages of this method are the considerable reduction of analysis time (2 min for acquiring metabolic profiles in both positive and negative polarities), simple analytical strategy owing to instrumental simplicity and data processing requirements, and the possibility to avoid the problems related to LC column deterioration and LC system maintenance, which all increase the robustness of the analytical procedure and reduce associated costs (Chekmeneva et al., 2017).

Similarly, studies carried to compare the metabolism of N-methyl-2-aminoindane (NM2AI) *in silico* and *in vivo* using MetaSite™ software and subsequently verified the presence of metabolites in the blood, urine, and hair of mice after NM2AI administration, respectively. The results are in agreement with software prediction and experimental results in biological samples. They have used LC-HRMS to detect the metabolites of NM2AI in urine, blood, and hair samples (Mestria et al., 2020).

Group 1 carcinogens like 17β-estradiol (E2) is a major environmental pollution factor of attention. *Novosphingobium* sp. ES2-1 is an effective estrogen-degrading bacterium that was isolated from the activated sludge. HRMS combined with $^{13}C_3$-labeling was used to identify the metabolites of E2. Figure 2.6 shows the possible scheme of oxidative degradation of E2 by the bacterium (Li et al., 2020). Table 2.3 depicts the metabolic profiling of various molecules and drugs in human by HRMS.

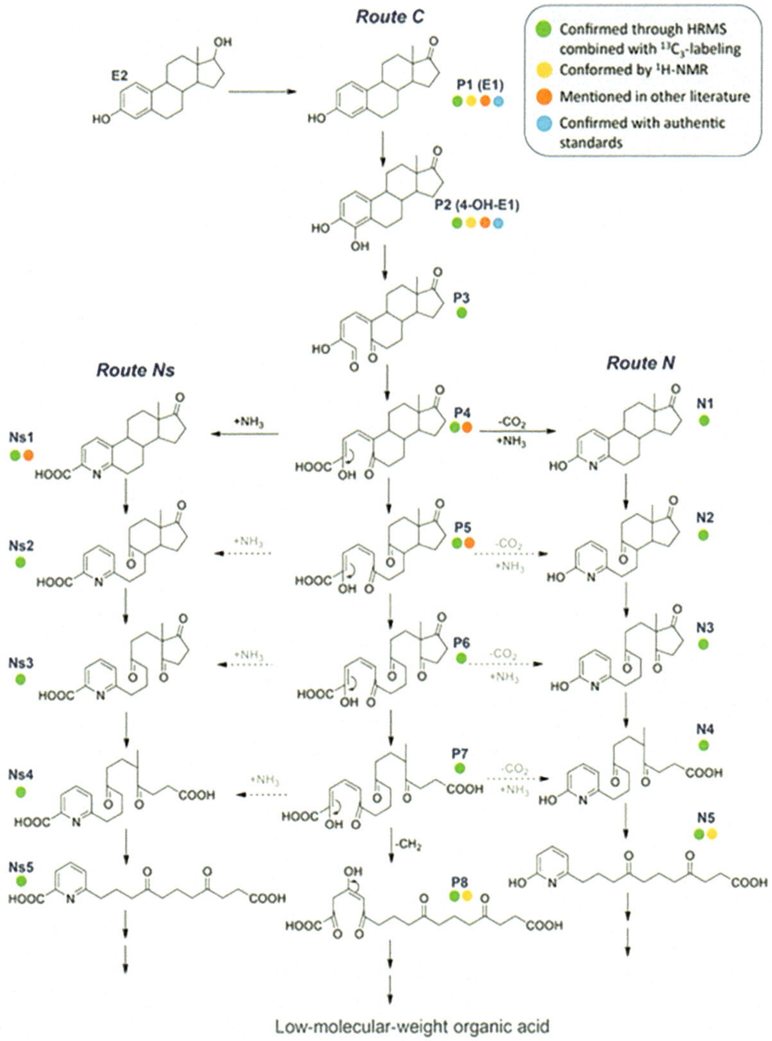

FIGURE 2.6 A possible pathway of oxidative degradation of 17β-estradiol (E2) by *Novosphingobium* sp. ES2-1. E1: Estrone; 4-OH-E1 (4-hydroxyestrone); P1-P8: products 1 to products 8 in route C; N1-N5: products 1 to products 5 in route N; Ns1-Ns5: products 1 to products 5 in route Ns.

Source: Reprinted with permission from Li et al,, 2020. © Elsevier.

TABLE 2.3 Metabolic Profiling of Various Molecules and Drugs in Human Identified by HRMS

Nature of Study	Purpose	HRMS Method	Molecules Identified	References
Human clinical trials using serum	Biomarkers of metastatic colorectal cancer	LC-HRMS	One endocannabinoid, two glycerophospholipids, and two sphingolipids	Martín-Blázquez et al. (2019)
In vitro by human liver microsomes fraction	Metabolites of vardenafil	LC-ESI-HRMS	Deethylated derivative of vardenafil along with new 9 molecules	Gawlik and Skibiński (2018)
Urinary metabolic profiling	Epidemiological investigations	Fit-for purpose direct infusion nanoelectrospray high-resolution mass spectrometry (DI-nESI-HRMS)	Metabolites including amino acids, alkaloids, organic acids	Chekmeneva et al. (2017)
Metabolites in the blood, urine, and hair of mice	Metabolites of N-methyl-2-aminoindane	LC-HRMS with Orbitrap mass detector	Seven main metabolites in the urine, as 2-aminoindane (2AI), two hydroxy-2AI and four hydroxy-N-methyl-2-aminoindane (NM2AI); one of the hydroxy-NM2AI and one of the hydroxy-2AI	Mestria et al. (2020)
Lung tissue of septic rats	Metabolic changes induced by cecal ligation and puncture and investigate the treatment effects of Xubijing injection (XBJ)	UHPLC-Q-Orbitrap-HRMS	Metabolic pathways included energy metabolism, amino metabolism, lipid metabolism, fatty acid metabolism and hormone metabolism. Among the 26-varied metabolites, 15 were significantly regulated after XBJ treatment	Xu et al. (2018)

TABLE 2.3 *(Continued)*

Nature of Study	Purpose	HRMS Method	Molecules Identified	References
In vivo in mice using serum, liver, kidney, urine	Metabolism of Triapine	LC-HRMS	Major metabolite as thiadiazole ring-closed oxidation product of Triapine and five monooxygenated metabolites	Pelivan et al. (2017)
In Arabidopsis T87 cultured cells	^{13}C- or 2H-labeled 2,4-dichlorophenoxyacetic acid (2,4-D)	LC-HRMS-MS	83 candidates for 2,4-D metabolites	Takahashi et al. (2018)
Human plasma	Detection of growth hormone-releasing hormones (GHRHs)	LC-HRMS/MS	Hormones including Geref (Sermorelin), CJC-1293, CJC-1295, and Egrifta (Tesamorelin) as well as two metabolites of Geref and CJC-1293	Knoop et al. (2016)
Human urine	Levamisole and its metabolite aminorex	LC-QTOF-HRMS and LC-QqQ-MS	Levamisole and aminorex detected in post-administration urine sample	Hess et al. (2013)
In vitro Phase I metabolic profile of N-(1-amino-1-oxo-3-phenylpropan-2-yl)-[1-(5-fluoropentyl)-1H-indole]-3-carboxamide	indole carboxamide synthetic cannabinoids	UHPLC-HRMS	A total of 10 metabolites were identified. Three monohydroxylated metabolites specific to PX-1 were reported for the first time.	Presley et al. (2020)

2.4 TRANSFORMATION AND DEGRADATION

It is essential to determine the degradation of drug molecules in pharmaceutical analysis, wastewater treatment and in microbial process. The awareness of the degradation pathways in water helps to control and monitor the formation of impurities and unknown molecules into the environment. HRMS is an important tool to identify the degradation mechanisms due to its high selectivity and sensitivity.

Pharmaceutical and cosmetic products are a major area for concern due to its effects on potential environmental and aquatic life after its discharge in huge quantity. Chlorination is an important process to treat the wastewater, but the information on the by-products are less. Hence Chen et al. established a HRMS method to identify the chlorinated by-products of three anti-inflammatory drugs, four parabens, bisphenol A, oxybenzone, and triclosan in comparison with untreated water products. The ethylparaben moiety has monochlorinated to ethyl 3-chloro-4-hydroxybenzoate and methylparaben has dichlorinated to methyl 3,5-dichloro-4-hydroxybenzoate as given in Figure 2.7. The results obtained by the structure elucidation by HRMS confirmed that chlorination could convert the drugs to more persistent, bioaccumulative, and toxic transformation products (TP). Thus the chlorination may pose increased risks to aquatic organisms (Chen et al., 2018).

Similarly, the transformation of the antidepressant drug venlafaxine (VFX) and its major metabolite O-desmethylvenlafaxine (DVFX) in the presence of UV/H_2O_2 were studied. The TP were studied using HRMS through UHPLC-linear ion trap high resolution Orbitrap instrument (LTQ-Orbitrap-MS). 11 and 6 transformational products were formed for VFX and DVFX, respectively, and the structure elucidation were carried out by MS^2 and MS^3 scans. Figure 2.8 shows the MS^2 and MS^3 spectra of transformational products with a base peak having molecular weight 310 (dihydroxilated-venlafaxine) of VFX. The fragment ion m/z 248 showed the simultaneous loss of the tertiary amine in C_{10} one of the two hydroxyl radicals, which could be located in C_{12} or C_{13} (García-Galán et al., 2016).

Recycling of low density polyethylene (LDPE) is used to make piping, trash bags, sheeting, and films for building and agricultural applications, composite lumber, and other products. But there are not many studies available to find the unknown substances in LDPE recycling. Yusa et al. developed an intelligent data acquisition with AcquireX™ linked to LC-Orbitrap Tribrid-HRMS (MS3) and coupled to Compound Discoverer data processing. Around 28 unknown molecules were identified consist of additives, mainly plasticizers, used in different plastic applications. Figure 2.9 gives the spectra of bis (2-ethylhexyl) adipate and octyl decyl phthalate which will fall in these molecules (Yusà et al., 2020).

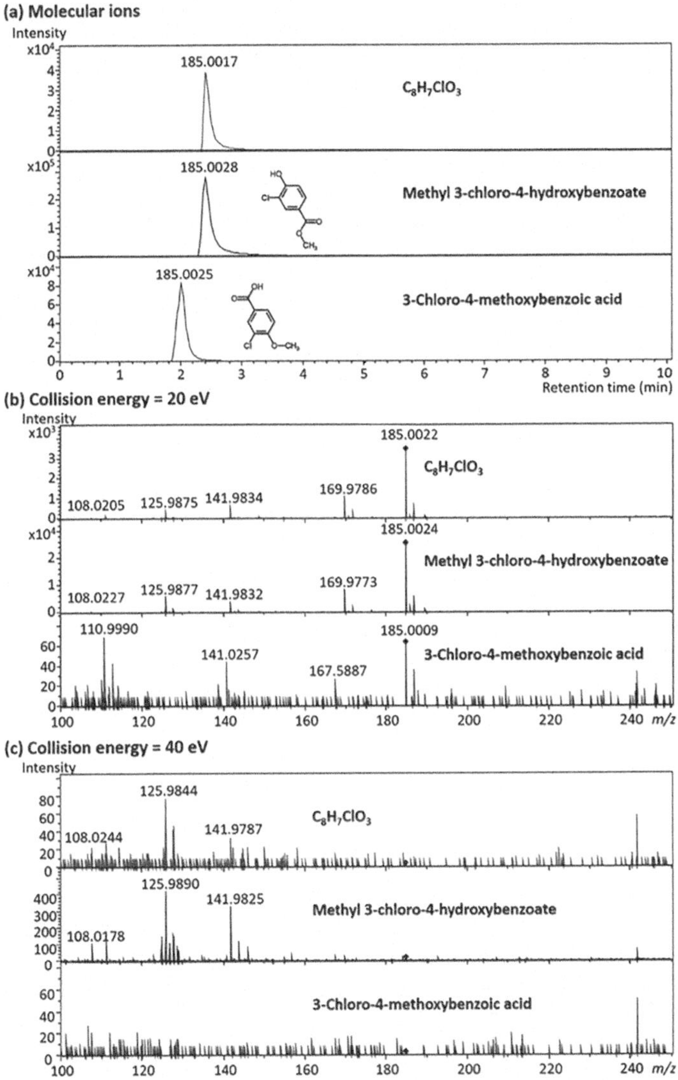

FIGURE 2.7 The retention times of the extracted molecular ion (a) and the fragment patterns at both; (b) high; and (c) low collision energies of the monochlorinated TP of methylparaben compared with those of the methyl 3-chloro-4-hydroxybenzoate.

Source: Reprinted with permission from Chen et al., 2018. © Elsevier.

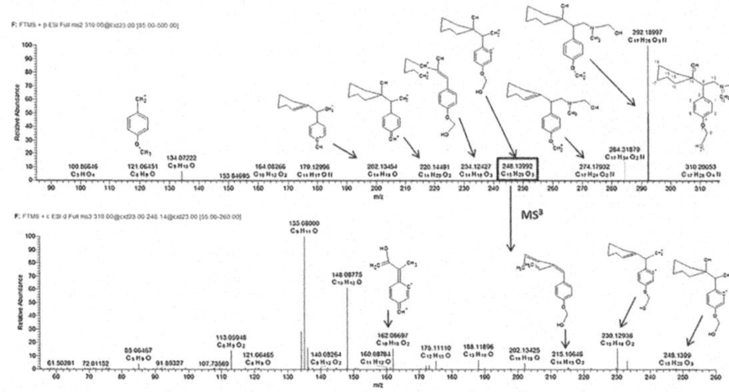

FIGURE 2.8 MS^2 and MS^3 spectra of transformational product having molecular weight 310 of VFX.

Source: Reprinted with permission from García-Galán et al., 2016. © Elsevier.

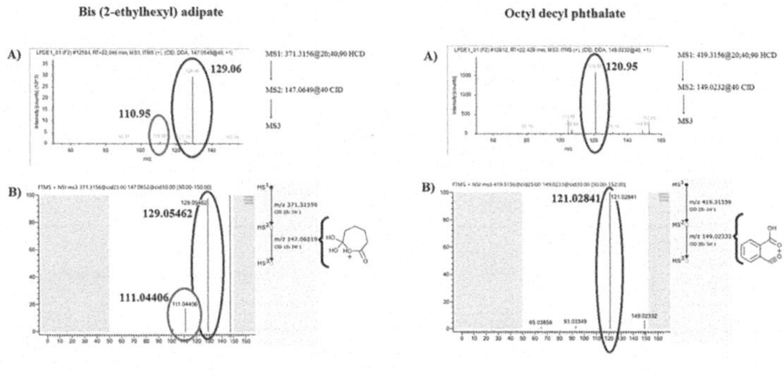

FIGURE 2.9 (A) Experimental MS^3 spectra; and (B) mzCloud MS^3 spectra for bis (2-ethylhexyl) adipate and octyl decyl phthalate.

Source: Reprinted with permission from Yusà et al., 2020. © Elsevier.

During the course of waste management, it is an important concern to identify the unknown materials, in order to decide whether they have to reuse or dispose, and how to do the same. This consideration find its heights when it comes to the identification of unknown chemicals or chemical weapons and their residues, both from past wars and armament production, and from the places of terrorist activities. Chen et al. present a description and application of a scheme to identify and quantify the unknowns from military wastes (Chen et al., 2004). Similarly, wastes from clinical laboratories, nuclear plants, chemical industries, etc., also need attention.

Table 2.4 depicts some of the degradation molecules identified by HRMS method from different molecules and drugs.

TABLE 2.4 The Degradation of Molecules Identified by HRMS Method from Different Molecules and Drugs

Nature of Study	Purpose	HRMS Method	Molecules Identified	References
Contaminants	Contaminants in a wastewater treatment plant effluent	LC-HRMS	New Six suspected and five non-target chemicals were identified and pollutants such as the herbicide clomazone, benzophenone-4, and benzothiazole	Hug et al. (2014)
Reduction of bacterial resistance	Ozonation of ciprofloxacin in water	HPLC-UV-HRMS	Reaction pathways are proposed starting with: (i) degradation at the piperazinyl substituent; (ii) degradation at the quinolone moiety with formation of isatin analogs; and (iii) degradation at the quinolone moiety with formation of anthranilic acid analogs.	DeWitte et al. (2008)
Reduction of drug resistance and increase degradation	Degradation kinetics of sulfamethoxazole (SMX) in an Fe(II)-activated persulfate	UHPLC-HRMS/MS	The S–N bond cleavage product, amino oxidation product and SO_2 extrusion product were detected in the Fe(II)-activated PS system	Luo et al. (2019)
Wastewater treatment	Degradation of polycaprolactone diol in aerobic bacteria	LC-HRMS	Ten transformation products were detected, 5 of them were the result of ester hydrolysis forming caprolactone oligomers while the other series corresponded to formation of polycaprolactone chain with a terminal diethylene glycol, likewise formed by ester hydrolysis	Rivas et al. (2017)
Wastewater treatment	Hydrolysis, photodegradation (UV irradiation) and chlorination experiments of duloxetine	UHPLC-ESI(+)-HRMS/MS	The parent compound was completely degraded after 30 min in photo-degradation and after 24 hr. in chlorination.	Osawa et al. (2019)

TABLE 2.4 *(Continued)*

Nature of Study	Purpose	HRMS Method	Molecules Identified	References
Wastewater treatment	Malachite green degradation by electrochemical method	HRMS	The intermediate of N-de-methylated derivatives of dye, i.e., N,N-dimethyl-N'-methyl-4,4'-Diaminotriphenylcarbenium; N-methyl-N'-methyl-4,4'-Diaminotriphenylcarbenium; N,N-dimethyl-4,4'-diaminotriphenylcarbenium; N-methyl-4,4'-diaminotriphenylcarbenium; 4,4'-diaminotriphenylcarbenium; N,N,N,'N'–tetramethyl-4,4'-diaminodiphenylepoxide; N,N-dimethyl-4-aminodiphenylepoxide; 4,4'–bis-aminobenzophenone	Singh et al. (2013)
Water treatment	Estrogen degradation pathway in activated sludge	UPLC-ESI-HRMS	Production of pyridinestrone acid and two A/B-ring cleavage products which are characteristic of the 4,5-seco pathway	Chen et al. (2018)
Comparison of electrochemical (EC) oxidation and natural microbial processes	Transformation products of carbamazepine by fungus *Pleurotus ostreatus*	LC-HRMS	EC prefers the acridine pathway, whereas the fungal treatment favors the diOH-CBZ pathway.	Seiwert et al. (2015)
Decomposition of methotrexate and doxorubicin	Photolysis and heterogeneous photocatalysis using titanium dioxide	LC-HRMS	Doxorubicin involved (poli)hydroxylation and/or oxidation, or the detachment of the sugar moiety. Methotrexate involved decarboxylation or the molecule cleavage.	Calza et al. (2014)

2.5 IMPURITY PROFILING

Impurity profiling is an inevitable process during the synthesis of a medicinal product or drug, and it is the process that includes identification, classification, and measurement of identical and non-identical pharmaceutical contamination present on the drugs. It helps to differentiate the toxicity, control the process conditions, consistency, and stability of the pharmaceutical products. HRMS is an important tool for fast, easy, and reliable exposure of molecules for toxicological purposes.

Liao et al. identified a dimeric impurity of Ceftolozane, an antibiotic, by using LC/HRMS, HRMSMS, H/D exchange and 2D-NMR studies. This helped them to establish an optimized condition to control the impurities via controlling the moisture in the drug and drying process. The structure of the ceftolozane and the dimer impurity are given in Figure 2.10 and the HRMS spectrum of the impurity is given in Figure 2.11 (Liao et al., 2019).

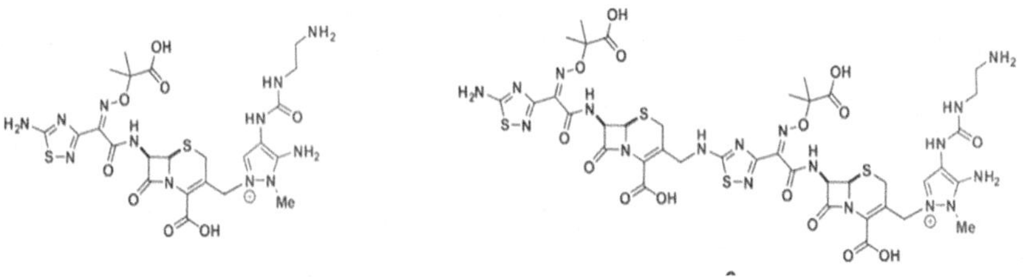

1. Ceftolozane **2. Dimeric impurity**

FIGURE 2.10 Structure of (1) ceftolozane; and (2) dimeric impurity. Major and minor fragments are shown in blue and red, respectively.

Source: Reprinted with permission from Liao et al., 2019. © Elsevier.

Pozzato et al. did a case study of acute constipation of Holstein cattle situated in Italy. Direct analysis in real-time (DART) coupled HRMS was used to identify the suspected hay batch and discovered that a distinct signal of m/z 507.2289 in the hay batch was thought to be associated with the digestive complications. LC-HRMS analysis suggested that the compound asperphenamate may cause acute toxicosis with depression, anorexia, ruminal, and intestinal impaction, with consequent drop in milk production, and possible deaths. The chromatogram and mass pattern of the Asperphenamate by ESI-HRMS/MS is given in Figure 2.12 (Pozzato et al., 2020).

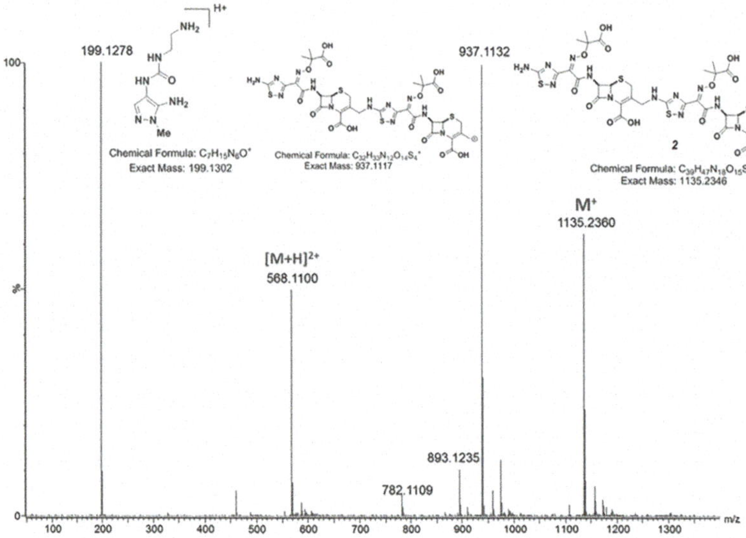

FIGURE 2.11 HRMS spectrum of impurity 2.

Source: Reprinted with permission from Liao et al., 2019. © Elsevier.

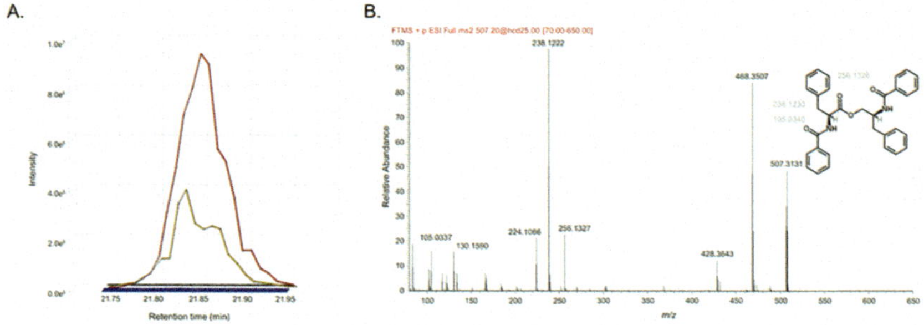

FIGURE 2.12 Asperphenamate confirmation and characterization. (A) Chromatographic peak of [Asperphenamate + H]$^+$ eluted at RT 28.1 (in red) in the suspected hay. No asperphenamate can be observed in regular hays (blue lines). (B) ESI-HRMS/MS spectrum of the [Asperphenamate + H]$^+$ of *m/z* 507.20 (NCE = 25). The cleavage sites are reported in the inset. (For interpretation of the references to color in this figure legend, the reader is referred to the web version of this chapter.).

Source: Reprinted with permission from Pozzato et al., 2020. © Elsevier.

Table 2.5 reveals some important studies on the impurity profiling, adulteration, and illegal pharmaceutical identification carried out by HRMS technique.

TABLE 2.5 The Impurity Profiling, Adulteration, and Illegal Pharmaceutical Identification Carried Out by HRMS Technique

Purpose	Target	HRMS Method	Molecules Identified	References
Impurities in synthetic thyroxine	Impurity profiling	UPLC-HRMS	Detection of 71 impurities, of which 47 have been unknown	Neu et al. (2013)
Adulteration in coffee	Oligosaccharide identification of soybeans and rice in ground coffee	UPLC-HRMS	17 oligosaccharide markers were determined	Cai et al. (2016)
Impurities in levothyroxine sodium	Impurity profiling	HPLC-HRMS/MS	Found two new classes of levothyroxine sodium impurities and 24 unknown molecules. 5 novel compounds were unambiguously identified	Ruggenthaler et al. (2017)
Illegal pharmaceuticals containing sildenafil	Illegal pharmaceuticals	LC-HRMS	23 elements and molecules like 5-chloroimidazosagatrizinone, imidazosagatriazinone, and aspartame were identified	Romolo et al. (2019)
Multiple classes of low-level impurities of phosphorothioate oligonucleotides	Synthesis impurities	LC-Fourier transform ion cyclotron resonance mass spectrometry (LC-FTMS)	3′-terminal phosphate monoester and 3′-terminal phosphorothioate monoester; incomplete backbone sulfurization and desulfurization products, high molecular weight impurities and chloral, isobutyryl, and N3 (2-cyanoethyl) adducts of the full length product.	Nikcevic et al. (2011)
Anticancer drug IIIM-290	Impurity profiling in pilot-scale batches	ESI-MS and HRMS	3 impurities were identified; rohitukine as m/z 305.32, rohitukine-N-oxide as m/z 322.13, IIIM-290-NO as m/z 478.08	Kumar et al. (2019)

TABLE 2.5 *(Continued)*

Purpose	Target	HRMS Method	Molecules Identified	References
Organic impurities in heroin	Forensic intelligence analysis	Direct analysis in real-time coupled with high-resolution mass spectrometry (DART-HRMS)	Niacin, paracetamol, ephedrine, caffeine, hydrocotarnine, lidocaine, morphine, codeine, thebaine, boldine, reticuline, papaverine, codamine, papaveraldine, laudanosine, heroin, N-acetylnorlaudanosine, 3,6-dimethoxy-8-(2-(N-methylacetamido)ethyl)phenanthren-4-yl acetate, N,O3,O6-triacetylnormorphine, noscapine, reticuline diacetate, narceine, and four unknowns were detected	Cui et al. (2019)
Cyclic peptide drug Arginine vasopressin (AVP)	Identification and quantification of structurally related peptide impurities	LC/HRMS/MS	3 deamidation products, ([Glu4]AVP, [Asp5]AVP, and AVP acid), two amino acid deletion impurities (des-Pro7-AVP and des-Gly9-AVP), one amino acid insertion impurity (endo-Gly10a-AVP), one end chain reaction product (N-acetyl-AVP), and one AVP isomer were detected	Wu et al. (2020a)
Impurity profiling of alfentanil hydrochloride	Degradation in stress conditions	LC/QTOF-MS/MS	17 related substances were detected in alfentanil HCl and its stressed samples, 9 were process-related substances and the other 8 were degradation products.	Wang et al. (2020)
Impurity profiling of the osteoporosis drug; calcitonin salmon	Structurally related peptide impurities	LC-HRMS	New impurities such as [7-dehydroalanine] calcitonin salmon, triple-sulfate-calcitonin salmon, [26-proline] calcitonin salmon, [14-Glutamic acid] calcitonin salmon, [20-glutamic acid] calcitonin salmon, [26-aspartic acid] calcitonin salmon, calcitonin salmon acid were observed	Wu et al. (2020b)

2.6 CONCLUSION

HR mass spectrometry techniques and their distinct properties make them an attractive platform for many modes of analyzes. Among them, the untargeted approach stay peculiar, as the strategies of HRFS runs encounter unexpected compounds of interest or unexpected concentrations. Their identification and confirmation, along with activity assignment is a great concern in many fields. The possibilities and outcomes of HRMS unknown identification strategies are discussed in the chapter, giving special emphasis for phytochemical biomarkers identification, pathway determination, Transformation, and degradation of compounds, impurity profiling, etc. A wise choice of HRMS instrumentation, sample preparation and data processing results in confident identification of the unknown of interest.

KEYWORDS

- **impurity profiling**
- **known unknowns**
- **liquid chromatography**
- **metabolic pathways**
- **phytochemicals**
- **unknown unknowns**

REFERENCES

Agüera, A., Martínez-Piernas, A. B., & Campos-Manas, M. C., (2017). Analytical strategies used in HRMS. In: Romero-González, R., & Frenich, A. G., (eds.), *Applications in High Resolution Mass Spectrometry* (pp. 59–82). Elsevier.

Bade, R., Bijlsma, L., Sancho, J. V., & Hernández, F., (2015). Critical evaluation of a simple retention time predictor based on LogKow as a complementary tool in the identification of emerging contaminants in water. *Talanta, 139*, 143–149.

Bastos, K. X., Dias, C. N., Nascimento, Y. M., Da Silva, M. S., Langassner, S. M. Z., Wessjohann, L. A., & Tavares, J. F., (2017). Identification of phenolic compounds from *Hancornia speciosa* (*Apocynaceae*) leaves by UHPLC orbitrap-HRMS. *Molecules, 22*, 143.

Cai, T., Ting, H., & Jin-Lan, Z., (2016). Novel identification strategy for ground coffee adulteration based on UPLC–HRMS oligosaccharide profiling. *Food Chem., 190*, 1046–1049.

Calza, P., Medana, C., Sarro, M., Rosato, V., Aigotti, R., Baiocchi, C., & Minero, C., (2014). Photocatalytic degradation of selected anticancer drugs and identification of their transformation products in water by liquid chromatography–high resolution mass spectrometry. *J. Chromatogr. A., 1362*, 135–144.

Chekmeneva, E., Dos Santos, C. G., Chan, Q., Wijeyesekera, A., Tin, A., Young, J. H., Elliott, P., et al., (2017). Optimization and application of direct infusion nanoelectrospray HRMS method for large-scale urinary metabolic phenotyping in molecular epidemiology. *J. Proteome Res., 16*, 1646–1658.

Chen, T. H., (2004). Identification of unknown solid waste. In: Twardowska, I., Allen, H. E., Kettrup, A. F., & Lacy, W. J., (eds.), *Waste Management Series* (Vol. 4, pp. 465–484). Elsevier: B.V.

Chen, W. L., Cheng, J. Y., & Lin, X. Q., (2018). Systematic screening and identification of the chlorinated transformation products of aromatic pharmaceuticals and personal care products using high-resolution mass spectrometry. *Sci. Total Environ., 637*, 253–263.

Chen, Y. L., Fu, H. Y., Lee, T. H., Shih, C. J., Huang, L., Wang, Y. S., Ismail, W., & Chiang, Y. R., (2018). Estrogen degraders and estrogen degradation pathway identified in an activated sludge. *Appl. Environ. Microbiol., 84*, e00001–18.

Cui, X., Lian, R., Chen, J., Ni, C., Liang, C., Chen, G., & Zhang, Y., (2019). Source identification of heroin by rapid detection of organic impurities using direct analysis in real-time with high-resolution mass spectrometry and multivariate statistical analysis. *Microchem. J., 147*, 121–126.

De Cássia, L. L. R., Kongstad, K. T, Kato, L., José, D. S. M., Franzyk, H., & Staerk, D., (2018). High-resolution PTP1B inhibition profiling combined with HPLC-HRMS-SPE-NMR for identification of PTP1B inhibitors from *Miconia albicans*. *Molecules, 23*, 1755.

DeWitte, B., Dewulf, J., Demeestere, K., Van De, V. V., De Wispelaere, P., & Van, L. H., (2008). Ozonation of ciprofloxacin in water: HRMS identification of reaction products and pathways. *Environ. Sci. Technol., 42*, 4889–4895.

García-Galán, M. J., Anfruns, A., Gonzalez-Olmos, R., Rodríguez-Mozaz, S., & Comas, J., (2016). UV/H_2O_2 degradation of the antidepressants venlafaxine and O-desmethylvenlafaxine: Elucidation of their transformation pathway and environmental fate. *J. Hazard. Mater., 311*, 70–80.

Gawlik, M., & Skibiński, R., (2018). Identification of new metabolites of vardenafil with the use of HLM and photochemical methods by LC-ESI-HRMS combined with multivariate chemometric analysis. *Int. J. Mass Spectrom., 433*, 55–60.

Hess, C., Ritke, N., Broecker, S., Madea, B., & Musshoff, F., (2013). Metabolism of levamisole and kinetics of levamisole and aminorex in urine by means of LC-QTOF-HRMS and LC-QqQ-MS. *Anal. Bioanal. Chem., 405*, 4077–4088.

Hu, M., Müller, E., Schymanski, E. L., Ruttkies, C., Schulze, T., Brack, W., & Krauss, M., (2018). Performance of combined fragmentation and retention prediction for the identification of organic micropollutants by LC-HRMS. *Anal. Bioanal. Chem., 410*, 1931–1941.

Hug, C., Ulrich, N., Schulze, T., Brack, W., & Krauss, M., (2014). Identification of novel micropollutants in wastewater by a combination of suspect and non-target screening. *Environ. Pollut., 184*, 25–32.

Kesting, J. R., Huang, J., & Sørensen, D., (2010). Identification of adulterants in a Chinese herbal medicine by LC–HRMS and LC–MS–SPE/NMR and comparative *in vivo* study with standards in a hypertensive rat model. *J. Pharm. Biomed. Anal., 51*, 705–711.

Kind, T., & Fiehn, O., (2006). Metabolomic database annotations via query of elemental compositions: Mass accuracy is insufficient even at less than1 ppm. *BMC Bioinformatics, 7*, 234.

Knoop, A., Thomas, A., Fichant, E., Delahaut, P., Schänzer, W., & Thevis, M., (2016). Qualitative identification of growth hormone-releasing hormones in human plasma by means of immunoaffinity purification and LC-HRMS/MS. *Anal. Bioanal. Chem., 408*, 3145–3153.

Kumar, V., Bhurta, D., Sharma, A., Kumar, P., Bharate, S. B., Vishwakarma, R. A., & Bharate, S. S., (2019). Impurity profiling of anticancer preclinical candidate, IIIM-290. *J. Pharm. Biomed., 166*, 1–5.

Li, J., Kuang, G., Chen, X., & Zeng, R., (2016). Identification of chemical composition of leaves and flowers from *Paeonia rockii* by UHPLC-Q-exactive orbitrap HRMS. *Molecules, 21*, 947.

Li, S., Liu, J., Williams, M. A., Ling, W., Sun, K., Lu, C., Gao, Y., & Waigi, M. G., (2020). Metabolism of 17β-estradiol by *Novosphingobium* sp. ES2-1 as probed via HRMS combined with 13C3-labeling. *J. Hazard. Mater., 389*, 121875.

Liao, J., Sheng, H., Saurí, J., Xiang, R., & Martin, G., (2019). Structural elucidation of a dimeric impurity in the process development of ceftolozane using LC/HRMS and 2D-NMR. *J. Pharm. Biomed. Anal., 174*, 242–247.

Lin, L. Z., Sun, J., Chen, P., Zhang, R. W., Fan, X. E., Li, L. W., & Harnly, J. M., (2014). Profiling of glucosinolates and flavonoids in *Rorippa indica* (Linn.) hiern. (Cruciferae) by UHPLC-PDA-ESI/HRMSn. *J. Agric. Food Chem., 62*, 6118–6129.

Little, J. L., Cleven, C. D., & Brown, S. D., (2011). Identification of "known unknowns" utilizing accurate mass data and chemical abstracts service databases. *J. Am. Soc. Mass Spectrom., 22,* 348–359.

Little, J. L., Williams, A. J., Pshenichnov, A., & Tkachenko, V., (2012). Identification of "known unknown" utilizing accurate mass data and ChemSpider. *J Am Soc Mass Spectrom., 23*, 1–7.

Liu, R., Zhao, Z., Dai, S., Che, X., & Liu, W., (2019). Identification and quantification of bioactive compounds in *Diaphragma juglandis* fructus by UHPLC-Q-orbitrap HRMS and UHPLC-MS/MS. *J. Agric Food Chem., 67*, 3811–3825.

Lobo, J. F. R., Vinther, J. M., Borges, R. M., & Staerk, D., (2016). High-resolution α-glucosidase inhibition profiling combined with HPLC-HRMS-SPE-NMR for identification of antidiabetic compounds in eremanthus crotonoides (Asteraceae). *Molecules, 21*, 782.

Luo, T., Wan, J., Ma, Y., Wang, Y., & Wan, Y., (2019). Sulfamethoxazole degradation by an Fe(ii)-activated persulfate process: Insight into the reactive sites, product identification and degradation pathways. *Environ. Sci.: Processes Impacts, 21*, 1560–1569.

Martín-Blázquez, A., Díaz, C., González-Flores, E., Franco-Rivas, D., Jiménez-Luna, C., Melguizo, C., Prados, J., et al., (2019). Untargeted LC-HRMS-based metabolomics to identify novel biomarkers of metastatic colorectal cancer. *Sci. Rep., 9*, 1–9.

Messaili, S., Colas, C., Fougère, L., & Destandau, E., (2020). Combination of molecular network and centrifugal partition chromatography fractionation for targeting and identifying Artemisia annua L. antioxidant compounds. *J. Chromatogr. A., 1615*, 460785.

Mestria, S., Odoardi, S., Federici, S., Bilel, S., Tirri, M., Marti, M., & Strano, R. S., (2020). Metabolism study of N-methyl 2-aminoindane (NM2AI) and determination of metabolites in biological samples by LC–HRMS. *J. Anal. Toxicol.*

Milman, B. L., (2011). *Chemical Identification and its Quality Assurance*. Springer, Berlin.

Milman, B. L., (2015). General principles of identification by mass spectrometry. *Trends Anal. Chem., 69*, 24–33.

Monga, G. K., Ghosal, A., & Ramanathan, D., (2019). To develop the method for UHPLC-HRMS to determine the antibacterial potential of a central American medicinal plant. *Separations, 6*, 37.

Neu, V., Bielow, C., Gostomski, I., Wintringer, R., Braun, R., Reinert, K., Schneider, P., et al., (2013). Rapid and comprehensive impurity profiling of synthetic thyroxine by ultrahigh-performance liquid chromatography–high-resolution mass spectrometry. *Anal. Chem., 85*, 3309–3317.

Nikcevic, I., Wyrzykiewicz, T. K., & Limbach, P. A., (2011). Detecting low-level synthesis impurities in modified phosphorothioate oligonucleotides using liquid chromatography–high resolution mass spectrometry. *Int. J. Mass Spectrom., 304*, 98–104.

Osawa, R. A., Carvalho, A. P., Monteiro, O. C., Oliveira, M. C., & Florêncio, M. H., (2019). Degradation of duloxetine: Identification of transformation products by UHPLC-ESI (+)-HRMS/MS, in silico toxicity and wastewater analysis. *J Environ Sci., 82*, 113–123.

Pelivan, K., Frensemeier, L., Karst, U., Koellensperger, G., Bielec, B., Hager, S., Heffeter, P., et al., (2017). Understanding the metabolism of the anticancer drug triapine: Electrochemical oxidation, microsomal incubation and *in vivo* analysis using LC-HRMS. *Analyst, 142*, 3165–3176.

Pozzato, N., Piva, E., Pallante, I., Bombana, D., Stella, R., Zanardello, C., Tata, A., & Piro, R., (2020). Rapid detection of asperphenamate in a hay batch associated with constipation and deaths in dairy cattle. The application of DART-HRMS to veterinary forensic toxicology. *Toxicon., 187*, 122–128.

Prakash, O., Baskaran, R., & Kudachikar, V. B., (2019). Characterization, quantification of free, esterified and bound phenolics in Kainth (*Pyrus pashia* Buch.-Ham. Ex D. Don) fruit pulp by UPLC-ESI-HRMS/MS and evaluation of their antioxidant activity. *Food Chem., 299*, 125114.

Presley, B. C., Logan, B. K., & Jansen-Varnum, S. A., (2020). Phase I metabolism of synthetic cannabinoid receptor agonist PX-1 (5F-APP-PICA) via incubation with human liver microsomes and UHPLC–HRMS. *Biomed. Chromatogr., 34*, e4786.

Rivas, D., Zonja, B., Eichhorn, P., Ginebreda, A., Pérez, S., & Barceló, D., (2017). Using MALDI-TOF MS imaging and LC-HRMS for the investigation of the degradation of polycaprolactone diol exposed to different wastewater treatments. *Anal. Bioanal. Chem., 409*, 5401–5411.

Rochat, B., (2017). Proposed confidence scale and ID score in the identification of known-unknown compounds using high resolution MS data. *J. Am. Soc. Mass Spectrom., 28*, 709–723.

Romolo, F. S., Salvini, A., Zelaschi, F., Oddone, M., Odoardi, S., Mestria, S., & Rossi, S. S., (2019). Instrumental neutron activation analysis (INAA) and liquid chromatography (LC) coupled to high resolution mass spectrometry (HRMS) characterization of sildenafil based products seized on the Italian illegal market. *Forensic Sci. Int., 1*, 126–136.

Ruggenthaler, M., Grass, J., Schuh, W., Huber, C. G., & Reischl, R. J., (2017). Levothyroxine sodium revisited: A wholistic structural elucidation approach of new impurities via HPLC-HRMS/MS, on-line H/D exchange, NMR spectroscopy and chemical synthesis. *J. Pharm. Biomed., 135*, 140–152.

Ruttkies, C., Schymanski, E. L., Wolf, S., Hollender, J., & Neumann, S., (2016). MetFrag relaunched: Incorporating strategies beyond in silico fragmentation. *J Cheminform., 8*, 1–16.

Schilling, C. H., Letscher, D., & Palsson, B. Ø., (2000). Theory for the systemic definition of metabolic pathways and their use in interpreting metabolic function from a pathway-oriented perspective. *J. Theor. Biol., 203*, 229–248.

Seiwert, B., Golan-Rozen, N., Weidauer, C., Riemenschneider, C., Chefetz, B., Hadar, Y., & Reemtsma, T., (2015). Electrochemistry combined with LC−HRMS: Elucidating transformation products of the recalcitrant pharmaceutical compound carbamazepine generated by the white-rot fungus *Pleurotus ostreatus. Environ. Sci. Technol., 49*, 12342−12350.

Shaikh, A., Tekale, S., Wagh, S., & Padul, M., (2020). Metabolite profiling of arginase inhibitor activity guided fraction of *Ficus religiosa* leaves by LC–HRMS. *Biomed. Chromatogr., 34*, e4966.

Silva, A. C. R., Da Silva, C. C., Garrett, R., & Rezende, C. M., (2020). Comprehensive lipid analysis of green Arabica coffee beans by LC-HRMS/MS. *Food Res. Int., 137*, 109727.

Singh, S., Srivastava, V. C., & Mall, I. D., (2013). Mechanism of dye degradation during electrochemical treatment. *J. Phys. Chem. C., 117*, 15229–15240.

Song, K., Oh, J. H., Lee, M. Y., Lee, S. G., & Ha, I. J., (2020). Molecular network-guided alkaloid profiling of aerial parts of *Papaver nudicaule* L. using LC-HRMS. *Molecules, 25*, 2636.

Takahashi, M., Izumi, Y., Iwahashi, F., Nakayama, Y., Iwakoshi, M., Nakao, M., Yamato, S., et al., (2018). Highly accurate detection and identification methodology of xenobiotic metabolites using stable isotope labeling, data mining techniques, and time-dependent profiling based on LC/HRMS/MS. *Anal. Chem., 90*, 9068–9076.

Tuenter, E., Foubert, K., Staerk, D., Apers, S., & Pieters, L., (2017). Isolation and structure elucidation of cyclopeptide alkaloids from *Ziziphus nummularia* and ziziphus spina-christi by HPLC-DAD-MS and HPLC-PDA-(HRMS)-SPE-NMR. *Phytochemistry, 138*, 163–169.

Tyrkkö, E., Pelander, A., & Ojanperä, I., (2012). Prediction of liquid chromatographic retention for differentiation of structural isomers. *Anal Chim Acta., 720*, 142–148.

Van, L. K. A., Nardin, T., Barker, D., Fedrizzi, B., Nicolini, G., & Larcher, R., (2020). A novel LC-HRMS method reveals cysteinyl and glutathionyl polysulfides in wine. *Talanta, 218*, 121105.

Vaniya, A., & Fiehn, O., (2015). Using fragmentation trees and mass spectral trees for identifying unknown compounds in metabolomics. *Trends Anal. Chem., 69*, 52–61.

Wang, Y., Xu, L., Lu, Y. T., Sun, F. M., Wu, D. F., Wang, J. Q., Di, B., et al., (2020). Impurity profiling of alfentanil hydrochloride by liquid chromatography/quadrupole time-of-flight high-resolution mass spectrometric techniques for drug enforcement. *Rapid Commun. Mass Spectrom., 34*, e8847.

Weber, R. J. M., & Viant, M. R., (2010). MI-pack: Increased confidence of metabolite identification in mass spectra by integrating accurate masses and metabolic pathways. *Chemometr. Intell. Lab. Syst., 104*, 75–82.

Wu, P., Li, M., Kan, Y., Wu, X., & Li, H., (2020). Impurities identification and quantification for calcitonin salmon by liquid chromatography-high resolution mass spectrometry. *J. Pharm. Biomed., 186*, 113271.b.

Wu, P., Ye, S., Li, M., Li, H., Kan, Y., & Yang, Z., (2020). Impurity identification and quantification for arginine vasopressin by liquid chromatography/high-resolution mass spectrometry. *Rapid Commun. Mass Spectrom., 34*, e8799.a.

Xu, T., Zhou, L., Shi, Y., Liu, L., Zuo, L., Jia, Q., Du, S., et al., (2018). Metabolomics approach in lung tissue of septic rats and the interventional effects of xuebijing injection using UHPLC-Q-orbitrap-HRMS. *J. Biochem., 164*, 427–435.

Yusà, V., López, A., Dualde, P., Pardo, O., Fochi, I., Pineda, A., & Coscolla, C., (2020). Analysis of unknowns in recycled LDPE plastic by LC-orbitrap tribrid HRMS using MS3 with an intelligent data acquisition mode. *Microchem. J., 158*, 105256.

Zhao, Y., Chen, M. X., Kongstad, K. T., Jäger, A. K., & Staerk, D., (2017). Potential of *Polygonum cuspidatum* root as an antidiabetic food: Dual high-resolution α-glucosidase and PTP1B inhibition profiling combined with HPLC-HRMS and NMR for identification of antidiabetic constituents. *J. Agric. Food Chem., 65*, 4421–4427.

CHAPTER 3

HRMS: The Cutting-Edge Technique in Clinical Laboratory

KARTHIK VARMA,[1] SREERAJ GOPI,[2] and JÓZEF T. HAPONIUK[2]

[1]R&D Center, Aurea Biolabs (P) Ltd., Kolenchery, Cochin – 682311, Ernakulam, Kerala, India, E-mail: ackvarmaapm@gmail.com

[2]Chemical Faculty, Gdansk University of Technology, Gdańsk, Poland, E-mail: sreerajgopi@yahoo.com (S. Gopi)

ABSTRACT

The uses of mass spectrometry have led researchers to overcome many hurdles in the field of analytical chemistry. For the identification of bioactive components, drugs, and metabolites, the use of high-end sophisticated instruments are required. High-resolution mass spectrometry (HRMS) is a highly sophisticated device with which we can identify and quantify many components by using the molecular mass. The use of HRMS plays a vital role in the field of clinical laboratory. The presence of toxicological drugs and substances in both clinical and forensic fields can be evaluated using HRMS. Other diverse applications of HRMS in the field of clinical laboratory include identification of molecular chain of vaccines, new psychoactive drugs (NPD), and molecular diagnostics.

High-Resolution Mass Spectrometry and Its Diverse Applications: Cutting-Edge Techniques and Instrumentation. Sreeraj Gopi, PhD, Sabu Thomas, PhD, Augustine Amalraj, PhD, & Shintu Jude, MSc (Editors)

3.1 INTRODUCTION

The uses of mass spectrometry (MS) have paved the way for the growing interest in the research field because of their versatile range of applications. The last few decades have witnessed the vast development of MS techniques in various analytical fields. The superiority of MS lies in its ability to identify molecules based on their mass.

High-resolution mass spectrometry (HRMS) not only offers higher resolution and better mass precision, but also higher sensitivity and larger dynamic range (Korfmacher et al., 2011). HRMS provides options for both full scan and product ion scans. By utilizing the full scan mode, information on metabolites, target ions and biomarkers can be quantitatively evaluated (Russell et al., 1999). In addition, full scan analysis of HRMS will help in reducing the method development time (Fung et al., 2011). The critical parameter involved in the quantitative analysis of HRMS is the selection of mass extraction window. The narrow mass extraction window explains the selectivity of the assay and is dependent on the resolution and sensitivity of the instrument (Kaufmann et al., 2010). Several studies have concentrated on the mass accuracy (MA) and mass extraction window (Rochat et al., 2012; Glauser et al., 2016).

For analysis in any MS, the instrumentation involves the generation of ions from the analyte solution. The most common and basic ionization techniques involve electrospray ionization (ESI), atmospheric pressure chemical ionization (APCI), atmospheric pressure photoionization (APPI), etc. ESI is most commonly employed for polar molecules and APCI for non-polar molecules. Another technique involves the generation of ions from the solid phase by matrix-assisted laser desorption ionization (MALDI). Ambient ionization techniques like desorption electrospray ionization (DESI), direct analysis in real-time (DART), etc., are also employed. For identification and quantification of molecules whose presence cannot be quantified using HPLC, the role of HRMS is dominant.

3.1.1 MULTIVARIED APPLICATION OF HRMS IN CLINICAL LABORATORY

The invention of HRMS has marked as of great importance in the field of various applications of clinical laboratories. The use of most modern instruments will surely have an impact on the research field. The chapter aims to present an overview on the wide use of HRMS in various domains of clinical laboratory (Table 3.1 and Figure 3.1).

TABLE 3.1 Multivariate Applications of HRMS in Clinical Laboratory

Application of HRMS	Analysis of Interest	References
Clinical toxicology	Quantification of drugs in urine, serum, and plasma	Sauvage et al. (2006)
	Quantification of poisons in urine, serum, and plasma	Cheng et al. (2006)
	Quantification of designer drugs in urine, serum, and plasma	Wu et al. (2006)
	Analysis of psychoactive drugs	Musshoff et al. (2006)
Molecular diagnostics	Analysis of testosterone hormones	Elmongy et al. (2020)
	Analysis of immunoglobulins	Handelsman et al. (2013)
	Determination of antibiotics	Cheng et al. (2017)
	Protein analysis	Han et al. (2008)
	Diagnosis of viral infectious diseases	Drake et al. (2011)
	RNA and DNA sequencing	Ragoussis et al. (2006)
Determination of vaccines	Determinations of sequencing of HIV virus	Haynes et al. (2015)
	Determination of sequencing of Ebola virus	Cazares et al. (2016)

3.2 CLINICAL AND TOXICOLOGICAL SCREENING

The aim of clinical toxicology is to obtain and analyze analytical results to support clinical diagnostics and treatment. Most of the conventional methods have the drawback of poor reproducibility and sensitivity, which are supposed to be the key factors involved in clinical toxicology analysis. Both gas chromatographic (GC) and liquid chromatographic (LC) mass spectrometric techniques are often employed in the clinical toxicology field. Both of the techniques have their own advantages and disadvantages in the field of clinical toxicology.

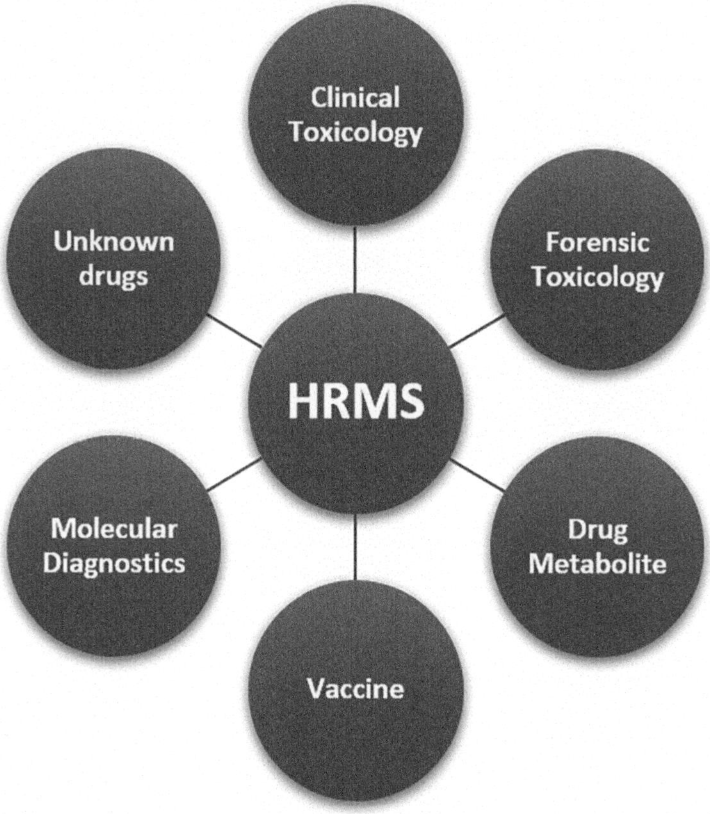

FIGURE 3.1 Multi-varied applications of HRMS in clinical laboratory.

The concept of using HRMS in clinical toxicology has been cited in literature because of their multiple advantages (Jiwan et al., 2011; Peters et al., 2011). In 2005, drug users on the US east coast admitted to the hospitals showed symptoms like hyperglycemia, palpitation, hypokalemia, etc., after consumption of heroin. When the toxicological studies were performed, evidence showing the use of opiates was only detected. However, upon further investigation, it was further confirmed the presence of Clenbuterol, a veterinary drug.

The common adulterants in cocaine (COC) can also be determined or quantified using HRMS techniques. The most common COC adulterants like diltiazem, hydroxyzine, methyl ephedrine, caffeine, lidocaine, benzocaine, procaine, phenacetin, levamisole, etc., are all subject of interest by the law enforcement agencies like US Food and Drug Administration (US FDA). A targeted GC MS or LC-MS/MS drug screening method can identify most of these pharmaceutical agents in COC samples (Cole et al., 2011; Fucci, 2007; Valentino et al., 2005; Casale et al., 2008). A study conducted by Laks et al. (2004) in the street drugs seized material without the presence of primary reference standards, was overcome by utilizing LC-TOF/MS in which the exact monoisotopic mass was determined. From the study, out of the 21 seized drug samples, more than 30 compounds were identified, which was further confirmed with reference to the reference standards available (Laks et al., 2004). Mass spectrometric methods can often be employed to determine the presence of drugs and their metabolites in blood, plasma, serum or urine. Mueller et al. (2005) discussed a screening procedure for 301 drugs in plasma and serum using a hybrid triple-quadrupole linear ion trap mass spectrometer (Q Trap) after liquid-liquid extraction (LLE) or mixed solid phase extraction (SPE). Sauvage et al. (2006) conducted a similar study, for normal screening purposes with the incorporation of new software features. The recovery of the 54 tested compounds ranged from 46% to more than 95%. The study found that the detection percentage of LC-MS/MS was 94%, GCMS 64% and LC-DAD 55% (Sauvage et al., 2006). Many studies concentrated on the procedures for the identification/quantification of drugs, poisons, and their metabolites in urine. Cheng et al. (2006) followed a classical approach for the identification and quantification of 19 drugs in urine after SPE. With the incorporation of a single point calibration curve, identification, and semi quantification of the drugs could be achieved with a 10 min programming including LOD, LOQ, matrix effect, etc. The limitation of the study included usage of an unlabeled drug Prazepam (Cheng et al., 2006).

An automated LC-MS/MS method was employed for the determination of five amphetamines (AP) and derived designer drugs in urine was evaluated with an overall run time of 20 minutes, but the study didn't consider the matrix effect, which remains as a constraint for its application (Wu et al., 2006). HRMS techniques can often be employed for the determination of low concentrations of benzodiazepines in the urine samples from drug-facilitated crime cases. The authors state that the adopted procedure is so sensitive, thereby reducing the run time, resources, and matrix load, which was further confirmed by an excretion study with some of the drugs showing long detection windows (Quintela et al., 2006). Identification of fentanyl derivatives or oral anti-diabetics in urine for doping control were performed using HRMS. The procedure involved enzymatic cleavage of conjugates followed by reversed-phase SPE for the fentanyls or LLE for the anti-diabetics, but the study does not focus on the matrix effect (Thevis et al., 2005). Similarly, determination of dialkyl phosphates, degradation products of organophosphate pesticides (OPPs) in urine using LLE reversed-phase separation was performed with the help of LC-MS/MS. The method was performed in the analysis of urine samples of a small cohort of non-exposed volunteers. The study stated a hint on the exposure to some frequently employed OPPs (Dulaurent et al., 2006). The importance of MS in the field of quantification of drugs and their metabolites in blood, plasma or serum is essential for the clinical toxicology reports. A fully automated turbulent flow MS procedure was developed for the identification of 13 anti-depressants and four of their pharmacologically active metabolites in serum for therapeutic drug (TD) monitoring. Turbulent flow methods involve direct injection of biological samples without any preliminary extraction and without any matrix interference. The authors stated the validity of the method with the inclusion of matrix effects and the authors checked external quality control (QC) samples with the new analytical procedure to check the reliability of the method (Sauvage et al., 2006). Juan et al. validated the applicability of LC-MS in the determination of anti-depressants in plasma after SPE. The study concentrated on the identification and quantification of the four common anti-depressants of the selective serotonin reuptake inhibitor (SSRI). However, the matrix lacks reliability, as the method was not validated for analyzing the matrix effects and selectivity (Juan et al., 2005).

Quintela et al. performed similar single stage LC-MS procedure for the determination of benzodiazepines in plasma and oral fluid (OF). The procedure was mainly developed for comparing the plasma and oral levels of midazolam and lormetazepam (Quintela et al., 2006). Similar studies were conducted by different authors for the determination of benzodiazepines in blood, urine, and hair samples (Bugey et al., 2006; Laloup et al., 2005). LC-MS/MS method was developed and validated for the detection of ephedrine-derived phenalkylamines in plasma samples (Beyer et al., 2007). Quantification of opioids and their metabolites can also be determined using a multi-analyte LC-MS/MS procedure, which included SPE and the method was validated (Musshoff et al., 2006). The metabolic profiling and quantification of antiarrhythmic drugs can be determined using LC-MS/MS. The method was developed using simple deproteinization (Li et al., 2007). Kristoffersen et al. developed a method for the determination of cardiovascular drugs in post-mortem whole blood after precipitation and mixed-mode solid-phase separation. The method was validated and robustness was determined (Kristoffersen et al., 2007). HRMS techniques can often be employed in the determination of psychoactive substances for the purpose of toxicological screenings.

3.3 MOLECULAR DIAGNOSTICS

High-end mass spectrometric instruments are often employed in the molecular diagnostics field. Mass spectrometric techniques are used in various disciplines of the molecular diagnostics because of their multiple advantages like multiplexing capacity and high analytical specificity. MS is the most commonly used technique in the clinical field for the determination of TD monitoring like immunosuppressant drugs, antiretroviral drugs, antiepileptic drugs, etc. It is also often employed for the determination of steroids like testosterone, androstenedione, 17-hydroxyprogesterone, aldosterone, cortisol, estrogen, etc. (Elmongy et al., 2020). Another application of HRMS relies in the field of endocrinology. It can be employed for the determination of biogenic amines, 5-hydroxyindoleacetic acid, insulin, thyroglobulin, etc. Mass spectrometric techniques are often employed in the field of immunology. Monoclonal immunoglobulins, serum-free light chains and IgG classes are also determined using mass spectrometric techniques (Handelsman et al., 2013). MALDI-ToF MS techniques are used in the molecular diagnostics field for the determination of hemoglobin A1c, insulin-like growth factor-1, multiplex biomarker panel of C reactive protein, serum amyloid A, fecal calprotectin and cystatin (Hattan et al., 2016). HRMS technique is also often employed in the microbiology field for the determination of antibiotics and anti-fungal drugs (Cheng et al., 2017). Mass spectrometric techniques are also employed in the field of anatomical pathology. Proteomic analysis can be performed in two different ways using MS techniques. It involves top-down proteomics, where the intact analysis of entire proteins are concentrated. However, the method lacks efficiency as most of the proteins involved in human diseases have almost similar masses, whereas the bottom-up method involves enzymatic cleavage of the protein complexes through the digestion of the peptidase (Han et al., 2008).

Mass spectrometric techniques play a predominant role in the detection of molecular diagnostics of viral infectious diseases (Drake et al., 2011). One of the most successful applications of MS is in the detection of pathogens involves the characterization of polymerase chain reaction (PCR) amplicons of the cystic fibrosis transmembrane conductance regulator gene. Similar techniques can often be employed in the fast detection of pathogens (Jurinke et al., 1996). Studies also acknowledged the use of MS in resolving the problems of RNA sequencing, resequencing, and methylation analysis (Ragoussis et al., 2006). Methylation, the key factor to understand the pathogenesis and carcinogenesis and has a significant role in gene regulation and the responses to infectious diseases. The first analytical assay for the precise quantification and analysis of methylation was based on MALDI-TOF MS (Vivekanandan et al., 2006). Genotyping is often employed to study the epidemiology and transmission of the pathogen responsible for the viral infection. MALDI-TOF MS has been deployed for viral typing, an important parameter in the genotyping. The technique included the extension of identification of major and subtype pathogen (Hong et al., 2008). Another successful assay for typing of influenza viruses was also performed using ESI-MS T5000 platform (Tang et al., 2004). MS-based techniques are often employed in post-genome sequencing, which plays a vital tool for viral typing, polymorphism analysis, DNA methylation, and RNA expression, etc. (Meyer et al., 2011). Viruses often generate genetic variations instantly or under other conditions, thereby resulting in a therapeutic failure. Automated MALDI TOF analyzes were employed for the detection of 60 different unknown mutations of the HBV genome.

3.4 ANALYSIS OF NEW PSYCHOACTIVE DRUGS (NPD)

Early 2000 have witnessed the use of new psychoactive drugs (NPD)/ substances (NPS) whose significance in medical history when compared to the known drugs were not established. A mass survey report conducted between the time period 2009 to 2017 have witnessed a mass increase in the number of NPD's, out of which the predominant ones were synthetic cannabinoids. The application of HRMS-QTOF in the detection of NPD's have gained much importance because of their relatively higher sensitivity to detect even trace amounts of substances. In a study conducted by

Kronstrand et al. (2014) LC-QTOF method for the detection of synthetic cannabinoid drugs were compared with a commercial immunoassay method and showed that the commercial method had the limitation of cross-reactivity to only a limited number of compounds (Kronstrand et al., 2014). Similar methods with slight modifications were performed for the detection of many other NPD of which the one conducted in the Australian states and territories had given a much insight to the use of NPD's. The use of DART which uses the principle of direct analysis of the sample by MS have gained importance nowadays. A method developed by Gwak et al. (2015) incorporates the use of ion mobility spectrometry (IMS) coupled with Q-TOF-MS utilizing a DART ionization source. With the incorporation of this method, the analysis of 35 NPS with an analysis time of less than 1 min was carried out (Gwak et al., 2015). Nie et al. (2016) developed a simple, direct analysis in real time mass spectrometry (DART-MS) method analysis of 11 NPSs including three categories of these substances present on the global market such as four cathinones, one phenylethylamine, and six synthetic cannabinoids. The 11 NPSs could be determined within the time of 0.5 min by DART-MS. Moreover, they could also be separated and determined within 5 min by the LC/QTOFMS method (Nie et al., 2016). The presence of NPD's in the hair can also be determined using high-end mass spectrometric instruments.

3.5 DETERMINATION OF VACCINES

High-resolution mass spectrometric techniques can also be used in the development of vaccines. The versatility of mass spectrometry to be used in vaccine development have made great impact in the development of new vaccines and even their toxicological aspects. The quest and necessity for the development of safe and effective vaccines can only be achieved using high-end instruments as sensitivity, repeatability, and reproducibility matters to a greater extend. Methods were developed on the hypothesis induce broadly neutralizing antibodies (bNAbs) that target the highly glycosylated HIV-1 virion-associated envelop (Haynes et al., 2015). Another global alarming virus that has resulted in the deaths of many people, especially in the African countries and throughout the world is EBOLA. Cazares et al. developed an HRMS based method for the quantitative determination of the viral envelope glycoprotein in Ebola virus-like particle vaccine preparations. The method mainly concentrated on the measurement of glycoprotein amounts in Ebola-like virus particles. The method employs an isotope dilution strategy using four target peptides from two regions of glycoproteins. Methods and conditions were optimized to ensure complete tryptic digestion of Ebola-like virus particles. The LC-HRMS method gave successful information on the quantification of glycoproteins with a lower quantification limit of 1 fmol and an average coefficient variation percent of 7.6%. Compared to the western blot values, the method exhibited a good linear relationship with survival in mice models (Cazares et al., 2016).

3.6 FORENSIC TOXICOLOGY

Forensic studies play a predominant role in the development of new regulations and governing bodies to have an insight on the use of many drugs. Governing bodies like scientific working group for the analysis of seized drugs (SWGDRUG) or Toxicology (SWGTOX) have all set minimum standards for the toxicological analysis and interpretation. They have categorized Mass spectra, as "Category A" analytical technique, providing the highest level of selectivity. As indicated in many of the toxicological standards, MS is regarded as a pioneer technique in the forensic account because of its selectivity and sensitivity. Both GC and LC-MS techniques are considered as "gold standards" in many of the forensic analysis.

3.7 IDENTIFICATION NEW ACTIVE COMPONENTS IN NATURAL PRODUCTS

HRMS plays a predominant role in the determination of new bioactive molecules whose molecular mass and structure is unknown. HRMS can often be employed for the quantification and qualification of new bioactive components. Studies suggest that more than 850 new drugs developed in the recent years are of natural origin (Guo et al., 2019). To detect the polyphenols in *Hypericum lanuginosum,* a targeted UHPLC-HRMS technique was employed for the determination of 21 polyphenols including phenolic acids, flavonoids, etc. The majority of the compounds were identified with reference standards, and some of them were identified using a fragmentation pattern (Mahomoodally et al., 2019). Another study, which concentrated on the determination of the composition of the cashew apple, used a non-targeted approach. The analysis used chemical fingerprinting in the near-infrared spectrometry to determine the bioactive components and degradation products. Several other natural products including galloylhexose I, digalloylhexoside I, hydroxybutanoic acid ethyl esterhexoside and other flavonoids were found using chemometrics evaluation of HRMS data. Several other glycosides are also evaluated using LC-HRMS, which was found as the most promising technique in the determination of glycosides (Marchetti et al., 2019). Cerulli et al. determined the occurrence of different bioactive components like giffonings and phenolics, which play a predominant role in the prevention of oxidative damage of human plasma lipids. Full scan with in-source collision-induced dissociation (CID) technique was employed to determine the bioactive components in *Cactus cladodes* (Cerulli et al., 2018). The analytical technique was found as a useful tool in the determination and bio-accessibility of polyphenols in both raw and cooked material and from the analysis; it paved a new pathway in the field of foodomics science (De Santiago et al., 2018). The HRMS technique can also be employed to determine and quantify the presence of several other bioactive components their metabolic pathway like peptides (Zenezini et al., 2016), carotenoids (Mi et al., 2019), terpenoids (Lü et al., 2019), glucosinolates (Capriotti et al., 2018), alkaloids, etc. (Rodríguez-Carrasco et al., 2018).

3.8 CONCLUSION

The use of high-end instruments in the field of research and various clinical field have always gained much attention due to their high degree of sensitivity and reliability. HRMS is widely used in various domains of clinical research like unknown drug identification, identification of drug metabolites, determination of vaccines, etc. In forensic toxicology, where every small quantity of drugs needs to be identified, HRMS will be a right tool. The analyte sample of interest can be determined even without any analytical standard. HRMS can be a versatile tool in various fields of research and product development in a cost effective manner.

KEYWORDS

- **clinical and forensic toxicology**
- **clinical laboratory**
- **high-resolution mass spectrometry**
- **identification of new drugs and intermediates**
- **mass spectrometry**
- **molecular diagnostics**

REFERENCES

Beyer, J., Peters, F. T., Kraemer, T., & Maurer, H. H., (2007). Detection and validated quantification of nine herbal phenalkylamines and methcathinone in human blood plasma by LC-MS/MS with electrospray ionization. *J. Mass Spectrom., 42*, 150–160.

Bugey, A., Rudaz, S., & Staub, C., (2006). A fast LC-APCI/MS method for analyzing benzodiazepines in whole blood using monolithic support. *J. Chromatogr, B Analyt. Technol. Biomed. Life Sci., 832*, 249–255.

Capriotti, A. L., Cavaliere, C., La Barbera, G., Montone, C. M., Piovesana, S., Zenezini, C. R., & Laganà, A., (2018). Chromatographic column evaluation for the untargeted profiling of glucosinolates in cauliflower by means of ultra-high performance liquid chromatography coupled to high resolution mass spectrometry. *Talanta, 179*, 792–802.

Casale, J. F., Corbel, E. M., & Hays, P. A., (2008). Identification of levamisole impurities found in cocaine exhibits. *Microgram J., 6*, 82–89.

Cazares, L. H., Ward, M. D., Brueggemann, E. E., Kenny, T., Demond, P., Mahone, C. R., Martins, K. O., et al., (2016). Development of a liquid chromatography high resolution mass spectrometry method for the quantitation of viral envelope glycoprotein in Ebola virus-like particle vaccine preparations. *Clin. Proteom., 13*, 18.

Cerulli, A., Napolitano, A., Masullo, M., Pizza, C., & Piacente, S., (2018). LC-ESI/LTQ orbitrap/MS/MS n analysis reveals diarylheptanoids and flavonol O -glycosides in fresh and roasted hazelnut (*Corylus avellana* cultivar "Tonda di Giffoni") *Nat. Prod. Commun., 13*, 1123–1126.

Cheng, W. C., Yau, T. S., Wong, M. K., Chan, L. P., & Mok, V. K., (2006). A high-throughput urinalysis of abused drugs based on a SPE-LC-MS/MS method coupled with an in-house developed post-analysis data treatment system. *Forensic Sci. Int., 162*, 95–107.

Cheng, W. M., Zhang, Q. L., Wu, Z. H., Zhang, Z. Y., Miao, Y. R., Peng, F., & Li, C. R., (2017). Identification and determination of myriocin in *Isaria cicadae* and its allies by LTQ-orbitrap-HRMS. *Mycology, 8*, 286–292.

Cole, C., Jones, L., McVeigh, J., Kicman, A., Syed, Q., & Bellis, M., (2011). Adulterants in illicit drugs: A review of empirical evidence. *Drug Test. Anal, 3*, 89–96.

De Santiago, E., Pereira-Caro, G., Moreno-Rojas, J. M., Cid, C., & De Peña, M. P., (2018). Digestibility of (poly)phenols and antioxidant activity in raw and cooked cactus cladodes (*Opuntia ficus-indica*). *J. Agric. Food. Chem, 6*, 5832–5844.

Drake, R. R., Boggs, S. R., & Drake, S. K., (2011). Pathogen identification using mass spectrometry in the clinical microbiology laboratory. *J. Mass. Spectrom, 46*, 1223–1232.

Dulaurent, S., Saint-Marcoux, F., Marquet, P., & Lachatre, G., (2006). Simultaneous determination of six dialkylphosphates in urine by liquid chromatography-tandem mass spectrometry. *J. Chromatogr. B. Anal. Technol. Biomed. Life. Sci., 831*, 223–229.

Elmongy, H., Masquelier, M., & Ericsson, M., (2020). Development and validation of a UHPLC-HRMS method for the simultaneous determination of the endogenous anabolic androgenic steroids in human serum. *J. Chromatogr. A, 1613*, 460686.

Fucci, N., (2007). Unusual adulterants in cocaine seizure on Italian clandestine market. *Forensic. Sci. Int., 172*, 2, 3.

Fung, E. N., Xia, Y. Q., Aubry, A. F.,' Zeng, J., Olah, T., & Jemal, M., (2011). Full scan high-resolution accurate mass spectrometry (HRMS) in regulated bioanalysis: LC-HRMS for the quantitation of prednisone and prednisolone in human plasma. *J. Chromatogr. B Anal. Technol. Biomed. Life Sci., 879*, 2919–2927.

Glauser, G., Grund, B., Gassner, A. L., Menin, L., Henry, H., Bromirski, M., Schütz, F., et al., (2016). Validation of the mass extraction window for quantitative methods using liquid chromatography high resolution mass spectrometry. *Anal. Chem., 88*, 3264–3271.

Guo, J., Lin, H., Wang, J., Lin, Y., Zhang, T., & Jiang, Z., (2019). Recent advances in bio-affinity chromatography for screening bioactive compounds from natural products. *J. Pharm. Biomed. Anal., 165*, 182–197.

Gwak, S., & Almirall, J. R., (2015). Rapid screening of 35 new psychoactive substances by ion mobility spectrometry (IMS) and direct analysis in real-time (DART) coupled to quadrupole time-of-flight mass spectrometry (QTOFMS). *Drug Test Anal., 7*, 884–893.

Han, X., Aslanian, A., & Yates, J. R., (2008). 3rd Mass spectrometry for proteomics. *Curr. Opin. Chem. Biol., 12*, 483–490.

Handelsman, D. J., & Wartofsky, L., (2013). Requirement for mass spectrometry sex steroid assays in the journal of clinical endocrinology and metabolism. *J. Clin. Endocrinol. Metab., 98*, 3971–3973.

Hattan, S. J., Parker, K. C., & Vestal, M. L., (2016). Analysis and quantitation of glycated hemoglobin by matrix-assisted laser desorption/ionization-time of flight mass spectrometry. *J. Am. Soc. Mass Spectrom., 27*, 532–541.

Haynes, B. F., (2015). New approaches to HIV vaccine development. *Curr. Opin. Immunol., 35*, 39–47.

Hong, S. P., Shin, S. K., Lee, E. H., Kim, E. O., Ji, S. I., Chung, H. J., Park, S. N., et al., (2008). High-resolution human papillomavirus genotyping by MALDI-TOF mass spectrometry. *Nat. Protoc., 3*, 1476–1484.

Jiwan, J. L., Wallemacq, P., & Hérent, M. F., (2011). HPLC-high resolution mass spectrometry in clinical laboratory? *Clin. Biochem., 44*, 136–147.

Juan, H., Zhiling, Z., & Huande, L., (2005). Simultaneous determination of fluoxetine, citalopram, paroxetine, venlafaxine in plasma by high performance liquid chromatography-electrospray ionization mass spectrometry (HPLC-MS/ESI). *J. Chromatogr. B. Analyt. Technol. Biomed. Life. Sci., 820*, 33–39.

Jurinke, C., Zöllner, B., Feucht, H. H., Jacob, A., Kirchhübel, J., Lüchow, A., Van, D. B. D., et al., (1996). Detection of hepatitis B virus DNA in serum samples via nested PCR and MALDI-TOF mass spectrometry. *Genet. Anal., 13*, 67–71.

Kaufmann, A., Butcher, P., Maden, K., Walker, S., & Widmer, M., (2010). Comprehensive comparison of liquid chromatography selectivity as provided by two types of liquid chromatography detectors (high-resolution mass spectrometry and tandem mass spectrometry): Where is the crossover point? *Anal. Chim. Acta, 673*, 60–72.

Korfmacher, W., (2011). High-resolution mass spectrometry will dramatically change our drug-discovery bioanalysis procedures. *Bioanalysis, 3*, 1169–1171.

Kristoffersen, L., Øiestad, E. L., Opdal, M. S., Krogh, M., Lundanes, E., & Christophersen, A. S., (2007). Simultaneous determination of 6 beta-blockers, 3 calcium-channel antagonists, 4 angiotensin-II antagonists and 1 antiarrhythmic drug in post-mortem whole blood by automated solid-phase extraction and liquid chromatography-mass spectrometry. Method development and robustness testing by experimental design. *J. Chromatogr. B. Analyt. Technol. Biomed. Life. Sci, 850*, 147–160.

Kronstrand, R., Brinkhagen, L., Birath-Karlsson, C., Roman, M., & Josefsson, M., (2014). LC-QTOF-MS as a superior strategy to immunoassay for the comprehensive analysis of synthetic cannabinoids in urine. *Anal. Bioanal. Chem., 406*, 3599–3609.

Laks, S., Pelander, A., Vuori, E., Ali-Tolppa, E., Sippola, E., & Ojanpera, I., (2004). Analysis of street drugs in seized material without primary reference standards. *Anal. Chem., 76*, 7375–7379.

Laloup, M., Ramirez, F. M. M., De Boeck, G., Wood, M., Maes, V., & Samyn, N., (2005). Validation of a liquid chromatography-tandem mass spectrometry method for the simultaneous determination of 26 benzodiazepines and metabolites, zolpidem and zopiclone, in blood, urine, and hair. *J. Anal. Toxicol., 29*, 616–626.

Li, S., Liu, G., Jia, J., Liu, Y., Pan, C., Yu, C., Cai, Y., & Ren, J., (2007). Simultaneous determination of ten antiarrhythmic drugs and a metabolite in human plasma by liquid chromatography-tandem mass spectrometry. *J. Chromatogr. B. Analyt. Technol. Biomed. Life. Sci., 847*, 174–181.

Lü, S., Zhao, S., Zhao, M., Guo, Y., Li, G., Yang, B., Wang, Q., & Kuang, H., (2019). Systematic screening and characterization of prototype constituents and metabolites of triterpenoid saponins of *Caulophyllum robustum* maxim using UPLC-LTQ Orbitrap MS after oral administration in rats. *J. Pharm. Biomed. Anal., 168*, 75–82.

Mahomoodally, M. F., Zengin, G., Zheleva-Dimitrova, D., Mollica, A., Stefanucci, A., Sinan, K. I., & Aumeeruddy, M. Z., (2019). Metabolomics profiling, biopharmaceutical properties of *Hypericum lanuginosum* extracts by *in vitro* and in silico approaches. *Ind. Crop Prod., 133*, 373–382.

Marchetti, L., Pellati, F., Graziosi, R., Brighenti, V., Pinetti, D., & Bertelli, D., (2019). Identification and determination of bioactive phenylpropanoid glycosides of Aloysia polystachya (*Griseb. Et moldenke*) by HPLC-MS. *J. Pharm. Biomed. Anal., 166*, 364–370.

Meyer, K., & Ueland, P. M., (2011). Use of matrix-assisted laser desorption/ionization time-of flight mass spectrometry for multiplex genotyping. *Adv. Clin. Chem., 53*, 1–29.

Mi, J., Jia, K. P., Balakrishna, A., Wang, J. Y., & Al-Babili, S., (2019). An LC-MS profiling method reveals a route for apocarotene glycosylation and shows its induction by high light stress in *Arabidopsis*. *Analyst, 144*, 1197–1204.

Mueller, C. A., Weinmann, W., Dresen, S., Schreiber, A., & Gergov, M., (2005). Development of a multi-target screening analysis for 301 drugs using a QTrap liquid chromatography/tandem mass spectrometry system and automated library searching. *Rapid. Commun. Mass. Spectrom., 19*, 1332–1338.

Musshoff, F., Trafkowski, J., Kuepper, U., & Madea, B., (2006). An automated and fully validated LC-MS/MS procedure for the simultaneous determination of 11 opioids used in palliative care, with 5 of their metabolites. *J. Mass Spectrom., 41*, 633–640.

Nie, H., Li, X., Hua, Z., Pan, W., Bai, Y., & Fu, X., (2016). Rapid screening and determination of 11 new psychoactive substances by direct analysis in real time mass spectrometry and liquid chromatography/quadrupole time-of-flight mass spectrometry. *Rapid. Commun. Mass Spectrom., 30*, 141–146.

Peters, F. T., (2011). Recent advances of liquid chromatography-(tandem) mass spectrometry in clinical and forensic toxicology. *Clin. Biochem., 44*, 54–65.

Ragoussis, J., Elvidge, G. P., Kaur, K., & Colella, S., (2006). Matrix-assisted laser desorption/ionization, time-of-flight mass spectrometry in genomics research. *PLoS Genet., 2*, e100.

Rochat, B., Kottelat, E., & McMullen, J., (2012). The future key role of LC-high resolution-MS analyses in clinical laboratories: A focus on quantification. *Bioanalysis, 4*, 2939–2958.

Rodríguez-Carrasco, Y., Gaspari, A., Graziani, G., Santini, A., & Ritieni, A., (2018). Fast analysis of polyphenols and alkaloids in cocoa-based products by ultra-high performance liquid chromatography and Orbitrap high resolution mass spectrometry (UHPLC-Q-Orbitrap-MS/MS). *Food Res. Int., 111*, 229–236.

Russell, D. H., & Edmondson, R. D., (1997). High-resolution mass spectrometry and accurate mass measurements with emphasis on the characterization of peptides and proteins by matrix-assisted laser desorption/ionization-time of–flight mass spectrometry. *J. Mass Spectrom, 32*, 263–276.

Sauvage, F. L., Gaulier, J. M., Lachatre, G., & Marquet, P., (2006). A fully automated turbulent-flow liquid chromatography-tandem mass spectrometry technique for monitoring antidepressants in human serum. *Ther. Drug Monit., 28*, 123–130.

Tang, Y. W., Lowery, K. S., Valsamakis, A., Schaefer, V. C., Chappell, J. D., White-Abell, J., Quinn, C. D., et al., (2013). Clinical accuracy of a PLEX-ID flu device for simultaneous detection and identification of influenza viruses A and B. *J. Clin. Microbiol, 51*, 40–44.

Thevis, M., Geyer, H., & Schanzer, W., (2005). Identification of oral antidiabetics and their metabolites in human urine by liquid chromatography/tandem mass spectrometry--a matter for doping control analysis. *Rapid. Commun. Mass. Spectrom, 19*, 928–936.

Valentino, A. M., & Fuentecilla, K., (2005). Levamisole: An analytical profile. *Microgram. J., 3*, 134–137.

Vivekanandan, P., Daniel, H. D., Kannangai, R., Martinez-Murillo, F., & Torbenson, M., (2010). Hepatitis B virus replication induces methylation of both host and viral DNA. *J. Virol., 84*, 4321–4329.

Wu, T. Y., & Fuh, M. R., (2005). Determination of amphetamine, methamphetamine, 3, 4 methylenedioxy amphetamine, 3,4-methylenedioxyethylamphetamine, and 3,4methylenedioxy methamphetamine in urine by online solid-phase extraction and ion-pairing liquid chromatography with detection by electrospray tandem mass spectrometry. *Rapid Commun Mass Spectrom., 19*, 775–780.

Zenezini, C. R., Capriotti, A. L., Cavaliere, C., La Barbera, G., Piovesana, S., & Laganà, A., (2016). Identification of three novel angiotens in converting enzyme inhibitory peptides derived from cauliflower by-products by multidimensional liquid chromatography and bioinformatics. *J. Funct. Foods, 27*, 262–267.

CHAPTER 4

Therapeutic Drug Designing, Discovery, and Development by HRMS

A. RAJESWARI,[1] E. JACKCINA STOBEL CHRISTY,[1] AUGUSTINE AMALRAJ,[2] and ANITHA PIUS[1]

[1]*Department of Chemistry, The Gandhigram Rural Institute – DU, Gandhigram, Dindigul – 624302, Tamil Nadu, India, Phone: +91-451-2452371 (O), Fax: +91-451-2454466 (O), E-mail: dranithapius@gmail.com (A. Pius)*

[2]*R&D Center, Aurea Biolabs (P) Ltd., Kolenchery, Cochin – 682311, Ernakulam, Kerala, India, E-mail: augustin14amal@gmail.com*

ABSTRACT

An efficient therapeutic drug (TD) has been found, for a particular disease, the therapeutic development process begins with the development of preclinical throughout increasingly complex and demanding phases of experimental studies to provision of authorization for promotion of marketing. The TD carrier systems have significantly used for the treatment and potential cure for various prolonged diseases, including diabetes, cancer, HIV, asthma, and drug addiction, whereas scientific efforts in drug designing and development are multidisciplinary. They are involving the physical, chemical, biological, pharma, engineering, and medical fields. In this chapter, we deal with TD designing, discovery, and development by high resolution mass spectrometry (HRMS).

High-Resolution Mass Spectrometry and Its Diverse Applications: Cutting-Edge Techniques and Instrumentation. Sreeraj Gopi, PhD, Sabu Thomas, PhD, Augustine Amalraj, PhD, & Shintu Jude, MSc (Editors)

4.1 INTRODUCTION

Therapeutics is a part of the medicinal field, particularly in the treatment of disease, art, and science of healing infections. In pharmacology, therapeutics is the use of drugs as well as the treatment of disease. The dosage of a therapeutic drug (TD) is related to the dose needed to treat a disease. Therapeutic agents are employed for healing different diseases caused by pathogens or malfunctioning organs.

A general practitioner has to personalize drug therapy of the patients in order to extent the optimal steadiness among therapeutic efficacy and the incidence of adverse effects. The identification of the goal is always difficult, because of each patient has different pharmacokinetic (PK) and pharmacodynamics (PD). When considering the PK and PD variability, the lack of control of drug concentration prevents the clinical response variability of the patients. In the 1960s, the estimation of the least drug concentrations found in biological liquids during drug treatment was accessible by utilizing new insightful procedures.

Conventional therapeutic candidates like small molecules are synthesized and administered without difficulty. A lot of these small molecules are not exact to their targets and may lead to side effects (Mócsai et al., 2014). Furthermore, causes for numerous diseases are the insufficiency of protein or enzyme. In this way, they can be dealt with utilizing biologically based treatments that can perceive a particular objective within crowded cells (Leader et al., 2008). Under the biologic conditions, a few macromolecules, like proteins and peptides are optimized to recognize specific targets (Roy et al., 2017). Consequently, they can dominate the inadequacy of small molecules.

Amino acid-based therapeutics like proteins, peptides, and peptidomimetics engineering works are recently receiving interest from pharmaceutical scientists (Farhadi et al., 2017, 2015a, 2015; Samish et al., 2011). Theoretical and experimental methods can predict the shape and folding of amino acid sequences and give an insight into how structure and function are encoded within the sequence. This type of identifications may be useful to interpret genomic information and life processes. Moreover, engineering of novel proteins or remodeling the existing proteins has opened the approaches to acquire novel biologic macromolecules with suitable therapeutic capabilities (Samish et al., 2011).

Therapeutic influence denotes to the responses afterward of any type of treatment, the results of which might be judged to be beneficial or favorable and true (Pagé et al., 2018; Kramer et al., 2003; Prima et al., 2008). An adversative consequence is the contrary and denotes to dangerous or undesired replies. Whatever constituents on the therapeutic impact versus a side effect may be a matter of each the situation of the nature and also the objectives of management. No inherent distinction separates therapeutic and unwanted effects; each response, behavioral/physiologic changes that occur as a response to the treatment strategy or agent. The scope of the treatment is to increase the therapeutic effects and decrease the side effects. It needs gratitude and quantification of the handling in various scopes. In a particular circumstance of targeted pharmaceutical medications, a combination of therapies is commonly required to realize the specified results (Stolker et al., 2004).

HRMS can be defined in different ways. One way to define HRMS is by resolving power. In the molecular weight range of small drug molecules (100–1,000 Da), quadrupole mass spectrometers are typically operated with unit mass resolution. Thus, a typical quadrupole MS can distinguish mass 500 and mass 501. This is suitable when the goal is SRM and a typical SRM transition example would be m/z 501.2 (precursor) to m/z 355.1 (product). On the other hand, if the goal is to use the exact mass of the compound of interest to separate it from endogenous isobaric compounds, then a higher resolving power is required. For decades HR mass spectrometers are available till the MS systems were designed to utilize as structural elucidation and/or metabolite identification (ID) tools. Quantitative applications were not well suited due to non-optimum scan speed, limited linear dynamic range and sub-optimum detection limits (Stolker et al., 2004). Thus, presently, triple quadrupole mass spectrometer is used by one group of scientists (that is the BA group) for quantitative analysis, while a second group of scientists (the metabolite ID group) use MS and/or HRMS systems for qualitative analysis. Now, the new HRMS systems are available which allow a new way instead of a paradigm shift to operating in the drug discovery BA arena.

4.2 THERAPEUTIC DRUG (TD) MOLECULES

Therapeutic molecules can be any chemical or synthetic metabolites synthesized by plants or microbes. Synthetic chemicals are modified to be drugs, or in some cases, they are accidentally turned out to have drug properties. The medical use of naturally available small drug molecules may be an accidental, since these compounds are not prepared by the microbes or plant with the projected perseverance of existing drugs. In another word, for the treatment of heart failure, the foxglove plant does not produce digitalis, rather, these naturally available metabolites are exploited by human being and repurposed for therapeutic utilize. In the present day drug discovery small molecule was a native out of and is a combination of above century ancient investigation approaches. The regular stages and periods of the drug discovery development, there is a strong general consensus (Nwaka and Ridley, 2003; Federsel, 2003). However, for drug discovery, there is no standardized method. They are exposed in its specific distinctive approach.

4.3 TARGET IDENTIFICATION AND VALIDATION

The accurate choice of the target drug for the selected sickness is the basic thing to decide the crucial achievement or disappointment of any drug discoveries. If a particular target is not in a pathway driving the fundamental origin of the sickness, or if the certain target disturbs additional metabolic methods in the body that effects viability then the drug discovery project will be a failure. The methods currently available to predict and effectively validate drug targets are limited. They are imprecise also, as is the information on the fundamental molecular etiology of numerous significant infections. These types of limitations are an important factor driving the high rate of failure in drug discovery. However, medicines are needed for patients, and new products are needed for biopharmaceutical companies and hence, the drug discovery enterprise carries on. The major directional conception in modern pharmacology is Paul Ehrlich's receptor theory (Figure 4.1).

Receptors are most commonly:

- The enzymes;
- Cellular switches;
- Signaling pathway elements.

While ancient medicines were discovered through good luck, the sector has modified in order that rather than finding a drug by fortune and so deciding how it performs, and the receptor targets are selected based on the scientific principles and medicines are sought which will perform on the preferred receptor targets. All infections have an original cause or molecular mechanism that creates the sickness called the disease molecular etiology. This plan can understand easily for infectious diseases. For instance, the molecular etiology of tuberculosis infections is due to the pathogenic bacterium Mycobacterium tuberculosis. That creates damage to the body and causes the symptoms of the disease. Arrhythmia is a major cause for heart disease, in which the heartbeat and rhythm of the beat are disturbed. That leads to a weakening and affect the function of the heart. The communication of a number of receptors and ion channels is a reason for the production of rhythm of the heartbeat. The rhythm of the heart can reestablish the heartbeat back to usual due to the drugs that perform on these regulators. Likewise, another cause of heart disease is high serum cholesterol. Hence, drugs that block the production of cholesterol in the body can lower cholesterol levels which lower the possibility of heart illness.

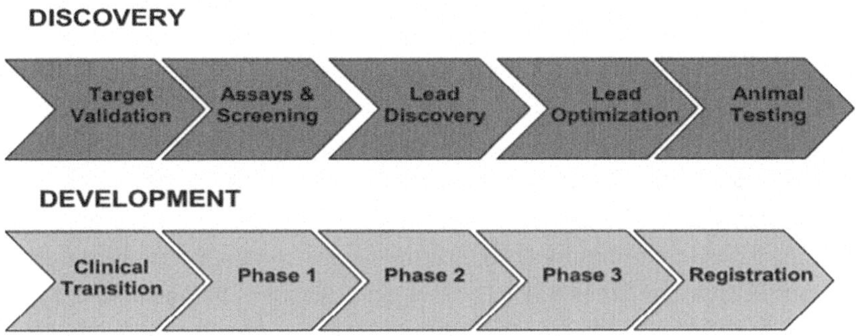

FIGURE 4.1 Drug discovery (therapeutic discovery and preclinical testing) and development (human clinical testing).

Subsequently, all that follows is influenced by the right choice of the target, abundant effort is placed interested in giving that sureness that the target has been properly chosen, which is called target validation. Target confirmation is unfortunately far-off from a particular effort. One overarching drawback is that for several sicknesses, the fundamental molecular etiology is not completely known with precision, and these are the illnesses for which new remedies are most urgently required (Tanzi and Bertram, 2001). The methods employed to validate hypotheses related to disease etiology are far from perfect because of their incapability to precisely model the disease in question. For such diseases, on trial hypotheses are considered to relate to what origins the sickness. While the drug inventor might initiate a basic analysis program to discover with certainty the cause behind the illness, this is an extended and strained out procedure with an undefined result (Tanzi and Bertram, 2001). So, a highly knowledgeable deduction will make by drug developers as to which of the challenging hypotheses are correct. Effective development of a drug will be one means to offer confirmation of the underlying hypothesis. For example, many scientists are working on Alzheimer's disease, by targeting a protein called beta-amyloid protein, considering the best treatment for Alzheimer's (Tanzi and Bertram, 2001). Others think the best approach would be to target a different protein called tau (Gotz, 2001). One group is maybe correct and the other wrong. Perhaps they are both right and the illness is affected by numerous protein dysfunctions or in a few cases of the infection are affected by the dysfunction of beta-amyloid protein and some cases are produced by the dysfunction of tau (Gotz, 2001). One more prospect is that both groups are incorrect and these protein dysfunctions are really an outcome of the illness. Here, the upstream from beta-amyloid and tau may be the actual cause. Although, the patients and their relatives needn't wait to find out. Affected people want researchers to effort in the drug development and the management in spite of the threat of failures. Depending on their best guess the drug designers move forward, whereas research groups are challenging with each other employing on beta-amyloid or tau. Even though, disease etiology is selected, they want to choose a target drug to counter the assumed reason for the sickness. Assume that the beta-amyloid causes Alzheimer's disease, then the next step is to arise with a tactic to block the damaging activities of beta-amyloid and to have high-quality fundamental indication associated with the validity of the method. Such proof is usually accumulated from various sources and approaches. One method is to initiate with an existing biochemical reagent which performs in the target in a nonselective approach. They investigate the component in a way that deducts out the non-particular actions of the chemical to try to identify the activity of a chosen drug. But, that type of reagents does not exist. One more approach is to use standard molecular biology methods. These methods create a prospective protein drug target in cells to up- or down-modulate via reagents that up- or down-regulate or even knockout (eliminate) the gene that encodes it. The principle is that simply building extra of the protein mimics a drug that stimulates the movement of the protein. Then constructing less of the protein mimics a drug that reduces the activity of the protein.

Another difficult issue is the scaffolding functions of many proteins in addition to their main action as an enzyme or receptor. Proteins generally appear in complexes with numerous other proteins. Varying the level of a protein can have effects on all the proteins present in the complex. Drugs are supposed to alter the enzyme or receptor activity only; at the same time the molecular biology techniques affect scaffolding activity and the enzyme or receptor activity. Another limitation is that, testing in the cells (*in vitro*) only comes close to a portion of what goes on in an intact animal (*in vivo*). Alternatively, gene knockouts or gene knock-ins can be produced in laboratory mice by molecular biology techniques. This is wherever the gene encoding the target of attention can be interrupted in the genome of a laboratory mouse (knockout). Otherwise an energetic form of the gene can be added into a mouse genome (knockin). A number of essential things about the target will be exposed from the description of the resultant mouse strain. The resultant knockout mouse strain will be lifeless or have about strong biological deficiency due to the essential function of the target. Thus the drug designer will be advised to leave from preventing a target that might result in toxic consequences. The knockout/knockin can be used in a mouse disease model. It is used to assess whether the genetic alteration changes of the course of the disease. Resulting that such a genetic modification, change the disease is very encouraging, though not conclusive because of a drug is specified solitary later the illness is identified. In contrast, the gene knockin or knockout is seen from the moment of beginning and definitely precedes the commencement of sickness indications in the animal model.

4.4 ASSAY DEVELOPMENT AND HIGH THROUGHPUT SCREENING

After the validation of therapeutic target, the next step is to find out the chemical compounds which act in the desired way on the target. This necessitates a screening test; a biochemical or biological assessment which will noticeably differentiate between active components that give the specified result, and inactive compounds that don't make a change to the target.

The screening assay should have the following characteristics:

- The assay must be comparatively low-cost, because thousands of compounds are usually verified in a distinctive screening campaign. If each assay like fully loaded for reagents, labor, lab, and overhead budgets 10 cents per analysis, expenditures for the screening test only will be around $20,000 excluding expenses for examining schemes and validation. As a common regulation of thumb, the cost of a single assay should be below $1 to facilitate the entire screening campaign will be below $200,000 to $300,000.
- The assay should be very sensitive. The investigator is looking aimed at lead of chemical, not a finalized drug in the project. One should anticipate that the unfinished chemical can lead to lower efficiency that may be better-quality by succeeding chemical preparation. Only a sensitive analysis could identify such crude and weak leads.
- The assay must be choosy for the action of importance. All biological examines are subject to focus on and find out compounds.

4.5 THERAPEUTIC ANTIBODIES

For the adaptive immune system, which is stimulated by a pathogen, antibodies are the primary effectors agents. Once stimulated, the immune system begins to create new antibodies which are physically bound to the pathogen and very selectively to the structural proteins which cause the pathogen construction. The antibodies will attach to the proteins of a definite pathogen. Normally, antibodies do not bind to the proteins which are portions of the own cells of the host. After binding, the antibodies regulate the immune system components to eliminate the pathogen. Once the disease resolves, if the same pathogen strikes again, the memory cells are recollected inside the immune arrangement to swiftly create the collection of antibodies. Vaccines are worked based on this mechanism. The immune system contains tremendously broad abilities. It is estimated that there are 10 billion different antibodies are made from human body. Each antibody is able to campaign to a specific feature of a foreign protein. Moreover, new antibodies are created once a pathogen infects an organism. Only the most common antibody features were identified because it is impossible to separate a specific antibody from this mixture. Before the year 1975, purified antibodies were only obtained from an unusual cancer termed as multiple myeloma. In multiple myeloma, a single antibody-producing cell convert proliferates and cancerous in the bone marrow. Enormous amounts of the single particular antibody were secreted by the cancer cells. In some cases, a huge amount of antibodies is produced and it spills out into the patient urine. The urine secreted proteins in this infection were known as Bence-Jones proteins by physicians who were not yet attentive to the nature of the spilled material. So, the whole field had to depend on these infrequent rarities of nature for their examinations. The only limitation is multiple myeloma cancers occur at random; there is no chance to identify what the antibody generates in the disease binds to. A monoclonal antibody method developed by Milstein through which pure individual antibodies subjected against any protein that be created in large quantity. Köhler and Milstein were honored with the Nobel Prize for the subject Medicine or Physiology in 1984 for their work (Rang, 2006). After those researchers and scientists started to think about the therapeutic applications of antibodies as it became technically feasible to make quantities of definite antibodies with a preferred set of chosen characteristics. As every new idea, the theory was also primarily convened with a number of negativities in the industry. Because antibodies are proteins, they would have to be controlled via injection. It may be several periods a day, a distinctive drawback comparative to traditional small molecule in orally active drugs. Once likened with traditional small molecule drugs, the production cost is higher for antibodies which potentially reduce profits. Then, therapeutic proteins manufacturing plants are expensive to build. Thus, it requires substantial capital investment. Eventually, industries started to escalate the encouraging features of healing antibodies. The preparation of antibodies is hard; at the same time, they are as well tough to reproduce, discouraging generic producers from appealing the market afterward patent expiration. When there are no pharmaceutical options and there are no available competing oral alternatives it could also be probable to arrive therapeutic expanses. Finally, antibodies have a tendency to live in the body for a very long period. The objective is to replace a naturally occurring protein with protein hormones or cytokines. In contrast, naturally occurring proteins inhibited and diminished their activity by the developed antibodies. It is clear that when a main protein is not there and wants to be replaced, for instance, in the case of insulin, assessing the therapeutic value of hindering a protein is much less straight forward. This way of identification of the target relatively look like the target credentials for small drug molecules (Nissim and Chernajovsk, 2008).

4.5.1 EFFICIENT THERAPEUTIC BIOMARKERS FOR BREAST CANCER

Small non-coding endogenous RNAs that contain 20–25 nucleotides are termed as microRNAs (miRNAs). They regulate gene expression at the post-transcriptional stage through sequence specific interactions with 3'-untranslated regions (UTRs) in miRNAs and inhibition of translation or degradation of miRNAs (Ambros, 2004; Kwak et al., 2010; Singh, 2017). In some essential biological methods such as metabolism, apoptosis, cell proliferation, etc., miRNAs play an important role, and are associated with carcinogenesis (Hwang and Mendell, 2007; Lu et al., 2005; Wang and Lee, 2010; Wu et al., 2010; Kohnehrouz et al., 2018). Aberrantly expressed miRNAs might work as oncogenes or tumor suppressor genes in cancers (Volinia et al., 2006; Yang et al., 2008; Wei et al., 2011; Liu et al., 2012).

NCI-60 Human Tumor Cell Lines from the DTP (Developmental Therapeutics Program) of NCI/NIH (National Cancer Institute/National Institutes of Health) includes nine types of cancer lines (Scudiero et al., 1988; Lorenzi et al., 2009). In the NCI-60 cell line panel is using multiple high-throughput screening data which includes DNA methylation, functional target analysis, compound screening data, gene expression data, protein analysis, data on changes in the DNA copy number and microRNA expression (Liu et al., 2010; Ikediobi et al., 2006; Nishizuka et al., 2003; Ehrich et al., 2008; Lee et al., 1994; Ma et al., 2018). This provides various chances at the genetic and molecular levels to detect specific pathways besides the mechanisms related to cancer.

Since miRNAs are involved in the pathogenesis of cancer, they can perform as potential molecular biomarkers. miRNAs may help to categorize cancers and determine their therapeutic response and also to discover novel targets (Gao et al., 2016; Li et al., 2014; Gaur et al., 2017). The anticancer effect of perifosine treatment was improved by inhibiting miR-720. The effect was decreased by miR-720 precursor treatment in breast cancer cell lines. Prevention of miR-887 can expand the anticancer influence of 5-Fu treatment and inhibition of miR-887 precursor treatment can decrease the anticancer influence of 5-Fu therapy, specifically in breast cancer cell lines. The current results give insight into the relationship between miRNA expression and their lineages. The approach used here can identify candidates with which to examine the resistance of drug and sensitive mechanisms in the future.

4.5.2 THERAPY FOR NEUROLOGICAL INFECTION

Quickly growing nontuberculous mycobacteria (NTM) are rarely been found through injuriousness concerning in bones or soft tissues, lungs, etc. (Wallace et al., 1989; Clegg et al., 1983; Galil et al., 1999; Winthrop et al., 2002; Talati et al., 2008). They can take place in contamination during surgery, contiguous spread from wounds and hematogenous injuriousness from a distant attention (Kanev and Sheehan, 2003; Kulkarni et al., 2001; Nau et al., 2010). Rarely these organisms may include the central nervous system (CNS) in addition. The NTM treatment is generally based on slight observational studies, by means of there are no proper clinical trials conducted in this area. Except a few case reports, there is not much literature available on CNS involving NTM. This lack of information, beside by the problems of permeation and adversative properties related with a long-term treatment creates the management really challenging. The factors to be carefully whereas selecting the treatment for such diseases can be understood by utilizing a condition of ventriculoperitoneal (VP) shunt infection with *Mycobacterium fortuitum*. A 13-years old female patient, affected by subsequent fossa glioma with bilateral VP shunt *in situ*, was identified to have acute appendicitis. She was subjected to laparoscopy and removed inflamed appendix. Around 15 days later the operation, she was admitted to the clinic with a very high fever and changed sensorium with convulsions. She had a Glasgow Coma Score of 9 (Eye-3, Verbal-2, Motor-4). On her abdomen, a large stable and moveable swelling was observed. Computerized tomography (CT) scans of the brain showed dilated ventricles. In the abdomen CT, observed a huge uniloculated cyst (pseudocyst) enfolding the shunt tips. A total leukocyte counts of glucose of 80 mg/dl, protein of 14 mg/dl and 280/μL (neutrophil 70%, lymphocyte 30%) were observed by cerebrospinal fluid (CSF) examination. Ziehl Neelsen staining of the CSF indicated multiple acid-fast bacilli (3+), and through line probe analysis and sequencing the development in culture was recognized as *Mycobacterium fortuitum*. The likely source of shunt disease in the case was the laparoscope employed in the appendectomy. They have theorized that the infection in the CSF was resulted from arising infection from the peritoneal conclusion of the VP shunt. All the shunt hardware were disconnected and a frontal Ommaya reservoir was implanted. Also, medicines have started. After the CSF became sterile the endoscopic third ventriculostomy was completed. Two months later, the patient was discharged with oral levofloxacin as well as linezolid and the oral treatment was stopped after 1½ years. In such cases, treatment necessitates suitable antimicrobials operative compared to the species of NTM affecting the infection. It is commonly recommended for the initial part of therapy to utilize at best one or two parenteral drugs, subsequently an elongated sequence of oral treatment. Four drugs containing regimen was chosen in this study, for the preliminary portion of the treatment, allowing for the participation of CNS and the cruelty of the situation and shortage of vulnerability analysis. For the same reason, the oral treatment with two drugs was extended past one year. The option was depended on the percentage susceptibility, adverse effect profile and penetration in CSF (Table 4.1). Because of the inherent property to form biofilms, the shunt hardware act as a source for colonization of the bacteria. It is necessary to remove of all such shunt material, in order to achieve a good outcome. CNS infection with NTM is an exceptional, however possibly thoughtful circumstance. Moreover, a high index of mistrust and primary identification, selecting the correct treatment for the right period is vital for a favorable conclusion.

4.5.3 NEW THERAPEUTIC DRUGS (TDS) FOR RARE DISEASES

Once a possible TD has been discovered, the method of emerging treatment for a specific disease starts with preclinical development and progress through progressively complex and challenging stages of medical testing to support approval for marketing. Necessary regulations are satisfied during the process of drug development which is driven by the requirement of the promoter of an innovative drug to reveal its protection and efficiency. Sometimes, public and non-profited or non-governmental organizations would take a product through this process, even though this work is expensive and risky, which has customarily been done within biotechnology and pharmaceutical companies (Field and Boat, 2010).

TABLE 4.1 CSF Penetration and Clinical Use of Different Classes of Antibiotics

Compound (CSF Penetration)	AUC_{CSF}/AUC_S Uninflamed or Mildly Description Inflamed Meninges	Strong Meningeal Inflammation	Relationship of CSF to MIC with Usual Doses	Compound with Broad Clinical Experience for CNS Infections	Description
Penicillins	0.02	0.2	CSF with uninflamed meninges close to the MICs for moderately susceptible bacteria	Penicillin G, ampicillin, amoxicillin	Low toxicity; daily dose can be increased up to 15–20 g
β-Lactamase inhibitors	0.07	0.1	CSF concentration with inflamed and uninflamed meninges below the concentration used *in vitro* for susceptibility testing (1–4 mg/liter)	Sulbactam	Little experience with *in vivo* activity in meningitis in humans; high dose sulbactam (up to 8 g/day) was used successfully to treat *Acinetobacter* meningitis
Chloramphenicol	0.6–0.7	0.6–0.7	CSF above the MIC for susceptible bacteria with uninflamed and inflamed meninges	Chloramphenicol	Bacteriostatic, risk of aplastic anemia; reserve antibiotic

About 10% of possible therapeutics that successfully passes preclinical development and reach the market. The cost for each (may range between $100 million and 1 billion), depends on the disease and other factors, including the cost of failed drugs. According to the reports from 50 largest pharmaceutical firms, approximately one in six new drugs which entered clinical testing, eventually received authorization for marketing. But based on the therapeutic class this rate varied widely and when compared to the drugs prepared by a company, the drug licensed into the company is slightly higher. In recent years the proportion has grown for orphan drug authorizations accounted by large pharmaceutical and medicinal industries. But the committees establish no examination of the attainment ratio definite to orphan drugs. The pharmaceutical and biotechnology industries generally motivation on prospective remedies with the maximum probability of creating an upright business profit, on the background concerns of the comparatively little probabilities of achievement and the great expenses of drug development. This clearly explains that prospective remedies for uncommon sicknesses, even the remedies for lifetime-threatening circumstances, have frequently suffered in the initial progress pipeline. In addition, drug development conventional approaches are often not possible for rare diseases that offers not only small markets, but also small populations for contribution in clinical studies (Field and Boat, 2010).

4.6 HRMS IN THERAPEUTIC DRUG (TD) DESIGNING, DISCOVERY, AND DEVELOPMENT

The goal of the current TD and drug metabolism (DM) science is to enable the discovery and development of safe and efficacious new molecular entities contained by a sensible interval and at a satisfactory rate. The impact of mass spectrometry (MS) technology on science is unequivocal (Korfmacher, 2005). Importance in utilizing high-resolution MS (HRMS) endures to magnify in the bioanalysis field in addition to further therapeutic claims (Xian et al., 2012; Ramanathan et al., 2011, 2012; Korfmacher, 2011; Ramanathan, 2013; Huang et al., 2013; Fung et al., 2013; King et al., 2014; Zelesky et al., 2013). In addition to TD and DM applications based on small molecules, HRMS is also being employed for big molecules including peptide and protein therapeutics (Ma and Chowdhury, 2013; Grondin et al., 2016; Wei et al., 2013; Morin et al., 2013; Dillen et al., 2012). Henry et al. described on a complete association of HRMS assays to SRM assays utilizing a triple quadrupole mass spectrometer for TDs analyzed in three various authorized SRM assays (Henry et al., 2012). The association established that HRMS delivered specificity, precision, accuracy, and sensitivity comparable to the SRM examines for the same drugs. One of the key benefits of HRMS is that SRM method development was not required for the assays. Furthermore, HRMS was identified to give beneficial sample troubleshooting info that would not have been accessible after the SRM assay of the same sample (Henry et al., 2012). Additionally, HRMS has the potential to give both quantitative and qualitative (quan/qual) information. In a study, King et al. (2014) concluded that recent HRMS systems can deliver bioanalytical data which are corresponding to information from triple quadrupole methods. But, there are processes that require to be implemented in order to effectively permit a (quan/qual) strategy for a TD and DM group (King et al., 2014). The progressions recommended by King et al. comprise the following:

- Confirm that the new quan/qual workflow does not compromise the data quality or throughput of the previous systems;
- Minimalize the variations in the operational performs required to appliance the quan/qual workflow;
- Confirm that personnel required to appliance the quan/qual methodology have familiarity in both parts;
- Scheme a perfect plan to explain which studies will be analyzed utilizing the quan/qual methodology;
- Confirm that the MS and data processing software packages are able to deliver the predictable outcomes in the period permissible for the method;
- Confirm that the local IT systems can handle the improved huge data documents that will be created by the MS system (King et al., 2014).

Moreover, the authors distinguished that MS dealers along with third party software dealers must to develop the quan/qual data software packages.

This special issue of bioanalysis focused on HRMS is a follow-up to the two special focus issues of bioanalysis on HRMS that were issued in 2012 and 2013. The 2012 publications concentrated predominantly on discovery bioanalysis plus metabolite identification applications of HRMS. The 2013 publications expanded the focus to include regulated bioanalysis, peptide quantitation and protein therapeutics, as well as editorials and opinions on various HRMS topics including using HRMS for forensic and clinical toxicology. The special issue of Bioanalysis on 2016, which focuses on HRMS, is a collection of articles that modernizes and magnifies the opportunity of how HRMS is being used by scientists engaged in various aspects of new drug discovery and drug development. This issue includes an opinion on HRMS as well as a commentary on HRMS (Ramanathan and Korfmacher, 2016; Rago and Negahban, 2016). The opinion discusses the rationales between the abbreviations/terminology HRMS and high-resolution accurate MS. The commentary provides readers some directions regarding the outsourcing ability of HRMS-based drug metabolism and pharmacokinetics (DMPK) studies and contract research organizations with HRMS capabilities. This is followed by a preliminary communication showing how HRMS was used in a metabolism study (Aratyn-Schaus and Ramanathan, 2016). The communication particulars in what way data-independent consecutive windowed acquisition of all hypothetical fragments-based HRMS technology is utilized to resolve a long-lasting metabolite formation enquiry. A perception on in what way, what time and why HRMS is executed in planning bioanalysis (Sturm, 2016). Some other researches magnify the claims of HRMS-based analyzes interested in antibody-drug conjugates and therapeutic proteins, respectively (Grafmuller et al., 2016; Weng, 2016).

The drug development procedures involve thorough examining and optimization of designated compounds to recognize the most effective drug. In order to examine the metabolism and to produce a product that is harmless, this testing is done *in vitro* and *in vivo* and has approved all regulatory necessities. Medications failure in medicinal practice is due in bigger part to two main reasons – if they do not work properly and if they are not safe. The two utmost significant concerns to report in drug creation procedures are to recognize a target and its validation. It may be a protein receptor which is combined with a condition of disease. Consequently, it is essential to recognize in what way the illness happens at the molecular, cellular, and genetic levels. When the target is recognized, then the next step is to find how the target performs a role in the illness progression. This is processed by testing the target against various recognized and innovative compounds to distinguish either distinct or numerous compounds that interact with the target and display either to nullify or to slow down the disease process.

Different drug development stages are:

- Invention of drug;
- Characterization of product;
- Formulation and development;
- Pharmacokinetics and drug disposition;
- testing and application of preclinical toxicology;
- Bioanalytical testing;
- Clinical trials;
- Regulatory review;
- The product or drug marketing;
- Post-marketing monitoring (Phase-IV) trials.

4.7 THERAPEUTIC DRUG MONITORING (TDM)

Therapeutic drug monitoring (TDM) is used as the medical practice of calculating particular drugs at selected intervals to sustain a constant attentiveness in the bloodstream of the patient, by which improving specific dosage schedules. It is not needed to employ TDM for all of the drugs. It is utilized for evaluating medicines with slight beneficial choices, drugs with noticeable PKs inconsistency, drugs for which target concentrations are hard to screen and medicines identified to cause healing and adversative effects. The procedure of TDM is supported on the assumption that there is a relationship between plasma and dose or blood drug concentration in addition to among concentration and healing effects. TDM starts when the medicine is first prescribed. It involves defining an initial dosage regimen suitable for the clinical condition and the patient characteristics such as age, organ function, concomitant drug therapy and weight. When concentration measurements are interpreting the sampling interval with respect to drug dose, history of dosage, responses of the patient and the preferred drug targets are to be considered. The aim of TDM is to use appropriate concentrations of hard to manage drugs to optimize medicinal results in subjects in different experimental trials.

Generally, TDM is recognized as the medical research laboratory assessments of a biochemical factor with appropriate medical interpretation and will directly affect drug prescribing process (Touw et al., 2005). On the other hand, TDM is the individualization of medication prescription by sustaining blood or plasma drug absorptions within a targeted therapeutic window or range (Birkett, 1997). TDM allows the assessment of the efficiency and safety of particular drugs in a variety of clinical settings by merging knowledge of pharmaceutics, PKs, and PD. The objective of this practice is to customize therapeutic procedures for optimum advantage for patients. By tradition, TDM comprises measuring the concentrations of drug in various biological fluids and identifying these concentrations related to clinical parameters. For the assessment of these interpretations, medicinal pharmacologists and pharmacists utilize principles of PKs. The rising of attentiveness of drug concentration-response associations, the charting of drug PKs characteristics, the initiation of high-throughput automation, and innovations in analytical expertise in turn encourage the development of clinical PK monitoring. Recent detonation of pharmacogenomic and pharmacogenetic investigation has been powered through the remarkable volume of genetic information produced by the human genome project (HGP). The HGP commenced its quest to record the comprehensive set of genetic information of the human genome, containing 3.2 billion base pairs coding up to 100,000 genes positioned on 23 pairs of chromosomes in 1990 (Carrico, 2000; Collins, 1999). While eventually conceived as a 15-year task, the HGP was really concluded by 2001 (Venter et al., 2001). Currently, developments in gene chip knowledge have accompanied in an innovative era of gene-based drugs and drug therapies.

4.8 NOVEL THERAPIES

The success and growth of the protein bio-therapeutics industry have prompted increased interest and investment into novel therapies. Research on stem cells as therapeutic agents is a growing field, and MS-based proteomics studies are frequently used to gain insight into the mechanisms of pluripotent cells and their interactions with their biological environments. MS is also a critical tool to monitor circulating levels of narrow therapeutic-range immunosuppressant drugs following stem cell or other transplant therapies (Hughes et al., 2012). Gene therapies are another emerging approach to disease treatment in which the biotherapeutic (a protein or peptide) is manufactured by a patient's own cells into which novel genetic material has been introduced. In this way, the cell can potentially regulate the expression of the gene and the production of the patient's 'medicine.' Again, MS analyzes are being used to detect and monitor the production of the desired biotherapeutic (Dadgari et al., 2010).

4.9 CONCLUSION

HRMS deconvolution has been proven to be an effective approach to quantify large TD designing, discovery, development, and monitoring. It can be an important complementary tool to the other bioanalytical platforms. In comparison to MS, the HRMS approach affords the capability to directly visualize the intact molecule at its high-level structure. Reasonably, TDM performs a vital role in the improvement of safe and actual therapeutic treatments and the individualization of these prescriptions.

KEYWORDS

- **3'-untranslated regions**
- **high resolution mass spectrometry**
- **immunosuppressant drugs**
- **nontuberculous mycobacteria**
- **pharmacodynamics**
- **therapeutic treatments**

REFERENCES

Ambros, V., (2004). The functions of animal microRNAs. *Nature, 431*, 350–355.

Aratyn-Schaus, Y., & Ramanathan, R., (2016). Advances in high resolution mass spectrometry and hepatocyte models solve a long-standing metabolism challenge: The loratadine story. *Bioanalysis, 8*, 1645–1662.

Birkett, D. J., (1997). Pharmacokinetics made easy: Therapeutic drug monitoring. *Aust. Prescr., 20*, 9–11.

Carrico, J. M., (2000). Human genome project and pharmacogenomics: Implication for pharmacy. *J. Am. Pharm. Assoc., 40*, 115–116.

Clegg, H. W., Foster, M. T., Sanders, W. E., & Baine, W. B., (1983). Infection due to organisms of the *Mycobacterium fortuitum* complex after augmentation mammaplasty: Clinical and epidemiologic features. *J. Infect. Dis., 147*, 427–433.

Collins, F. S., (1999). Shattuck lecture: Medical and societal consequences of the human genome project. *N. Engl. J. Med., 341*, 28–37.

Dadgari, J. M., Moore, R. E., Louie, K. A., Lee, T. D., & McMillan, M., (2010). Mass spectrometry measurement of a therapeutic peptide for use in multiple sclerosis. *Gene Ther., 17,* 709–712.

Dillen, L., Cools, W., Vereyken, L., Lorreyne, W., Huybrechts, T., De Vries, R., Ghobarah, H., & Cuyckens, F., (2012). Comparison of triple quadrupole and high-resolution TOF-MS for quantification of peptides. *Bioanalysis, 4,* 565–579.

Ehrich, M., Turner, J., Gibbs, P., Lipton, L., Giovanneti, M.,Cantor, C., & Van, D. B. D., (2008). Cytosine methylation profiling of cancer cell lines. *Proc. Natl. Acad. Sci. U S A., 105,* 4844–4849.

Farhadi, T., & Hashemian, S. M. R., (2017). Constructing novel chimeric DNA vaccine against Salmonella enterica based on SopB and GroEL proteins: An in silico approach. *Int. J. Pharm. Investig.,* 1–17.

Farhadi, T., Nezafat, N., & Ghasemi, Y., (2015). In silico phylogenetic analysis of Vibrio cholera isolates based on three housekeeping genes, *Int. J Comput. Biol. Drug Des., 8,* 62–74.

Farhadi, T., Nezafat, N., Ghasemi, Y., Karimi, Z., Hemmati, S., & Erfani, N., (2015a). Designing of complex multi-epitope peptide vaccine based on Omps of Klebsiella pneumoniae: An in silico approach. *Int. J. Pept. Res. Ther., 21*, 325–341.

Federsel, H., (2003). Asymmetry on large scale: The roadmap to stereoselective processes. *Nat. Rev. Drug Discov., 2*, 685–697.

Field, M. J., & Boat, T. F., (2010). Development of new therapeutic drugs and biologics for rare diseases. In: Field, M. J., & Boat, T. F., (eds.), *Rare Diseases and Orphan Products: Accelerating Research and Development*. National Academies Press, US.

Fung, E. N., Jemal, M., & Aubry, A. F., (2013). High-resolution MS in regulated bioanalysis: Where are we now and where do we go from here? *Bioanalysis, 5,* 1277–1284.

Galil, K., Miller, L. A., Yakrus, M. A., Wallace, Jr. R. J., Mosley, D. G., England, B., Huitt, G., et al., (1999). Abscesses due to mycobacterium abscessus linked to injection of unapproved alternative medication. *Emerg. Infect. Dis., 5,* 681–687.

Gao, X., Wu, Y., Yu, W., & Li, H., (2016). Identification of a seven-miRNA signature as prognostic biomarker for lung squamous cell carcinoma. *Oncotarget, 7*, 81670–81679.

Gaur, P., & Chaturvedi, A., (2017). Clustering and candidate motif detection in exosomal miRNAs by application of machine learning algorithms. *Interdiscip Sci., 3,* 1–9.

Gotz, J., (2001). Tau and transgenic animal models. *Brain Res. Rev., 35*, 266–286.

Grafmuller, L., Wei, C., Ramanathan, R., Barletta, F., Steenwyk, R., & Tweed, J., (2016). Unconjugated payload quantification and DAR characterization of antibody-drug conjugates using high resolution mass spectrometry (HRMS). *Bioanalysis, 8*, 1663–1678.

Grondin, A., Pillon, A., Vandenberghe, I., Guilbaud, N., Kruczynski, A., & Gomes, B., (2016). Discovery bioanalysis and *in vivo* pharmacology as an integrated process: A case study in oncology drug discovery. *Bioanalysis, 8*, 1481–1498.

Henry, H., Sobhi, H. R., Scheibner, O., Bromirski, M., Nimkar, S. B., & Rochat, B., (2012). Comparison between a high-resolution single-stage orbitrap and a triple quadrupole mass spectrometer for quantitative analyses of drugs. *Rapid Commun. Mass Spectrom., 26*, 499–509.

Huang, M. Q., Lin, Z. J., & Weng, W., (2013). Applications of high-resolution MS in bioanalysis. *Bioanalysis, 5,* 1269–1176.

Hughes, C., Radan, L., Chang, W. Y., Stanford, W. L., Betts, D. H., Postovit, L. M., & Lajoie, G. A., (2012). Mass spectrometry based proteomic analysis of the matrix microenvironment in pluripotent stem cell culture. *Mol. Cell. Proteomics, 11,* 1924–1936.

Hwang, H. W., & Mendell, J. T., (2007). MicroRNAs in cell proliferation, cell death and tumorigenesis. *Br. J. Cancer, 96,* 40–44.

Ikediobi, O. N., Davies, H., Bignell, G., Edkins, S., Stevens, C., O'Meara, S., Santarius, T., et al., (2006). Mutation analysis of 24 known cancer genes in the NCI-60 cell line set. *Mol. Cancer Ther., 5,* 2606–2612.

Kanev, P. M., & Sheehan, J. M., (2003). Reflections on shunt infection. *Pediatr. Neurosurg., 39,* 285–290.

King, L., Kotian, A., & Jairaj, M., (2014). Introduction of a routine Quan/qual approach into research DMPK: Experiences and evolving strategies. *Bioanalysis, 6,* 3337–3348.

Kohnehrouz, B. B., Bastami, M., & Nayeri, S., (2018). In silico identification of novel microRNAs and targets using EST analysis in *Allium cepa* L. *Interdiscip. Sci., 10,* 771–780.

Korfmacher, W. A., (2005). Principles and applications of LC-MS in new drug discovery. *Drug Discov. Today, 10,* 1357–1367.

Korfmacher, W., (2011). High-resolution mass spectrometry will dramatically change our drug-discovery bioanalysis procedures. *Bioanalysis, 3,* 1169–1171.

Kramer, T., (2003). Side effects and therapeutic effects. *MedGenMed: Medscape General Medicine, 5,* 28.

Kulkarni, A. V., Drake, J. M., & Lamberti-Pasculli, M., (2001). Cerebrospinal fluid shunt infection: A prospective study of risk factors. *J. Neurosurg., 94,* 195–201.

Kwak, P. B., Iwasaki, S., & Tomari, Y., (2010). The microRNA pathway and cancer. *Cancer Sci., 101,* 2309–2315.

Leader, B., Baca, Q. J., & Golan, D. E., (2008). Protein therapeutics: A summary and pharmacological classification. *Nat. Rev. Drug Discov., 7,* 21–39.

Lee, J. S., Paull, K., Alvarez, M., Hose, C., Monks, A., Grever, M., Fojo, A. T., & Bates, S. E., (1994). Rhodamine efflux patterns predict P-glycoprotein substrates in the national cancer institute drug screen. *Mol. Pharmacol., 46,* 627–638.

Li, X., Shi, Y., Yin, Z., Xue, X., & Zhou, B., (2014). An eight-miRNA signature as a potential biomarker for predicting survival in lung adenocarcinoma. *J. Transl. Med., 12,* 159–163.

Liu, H., D'Andrade, P., Fulmer-Smentek, S., Lorenzi, P., Kohn, K. W., Weinstein, J. N., Pommier, Y., & Reinhold, W. C., (2010). mRNA and microRNA expression profiles of the NCI-60 integrated with drug activities, *Mol. Cancer Ther., 9,* 1080–1091.

Liu, R., Chen, X., Du, Y., Yao, W., Shen, L., Wang, C., Hu, Z., et al., (2012). Serum microRNA expression profile as a biomarker in the diagnosis and prognosis of pancreatic cancer. *Clin. Chem., 58,* 610–618.

Lorenzi, P. L., Reinhold, W. C., Varma, S., Hutchinson, A. A., Pommier, Y., Chanock, S. J., & Weinstein, J. N., (2009). DNA fingerprinting of the NCI-60 cell line panel. *Mol. Cancer Ther., 8,* 713–724.

Lu, J., Getz, G., Miska, E. A., Alvarezsaavedra, E., Lamb, J., Peck, D., Sweetcordero, A., Ebert, B. L., et al., (2005). MicroRNA expression profiles classify human cancers. *Nature, 435*, 834–838.

Ma, Q., Song, J., Chen, Z., Li, L., Wang, B., & He, N., (2018). Identification of potential gene and microRNA biomarkers for colon cancer by an integrated bioinformatical approach. *Transl. Cancer Res., 7,* 17–29.

Ma, S., & Chowdhury, S. K., (2013). Data acquisition and data mining techniques for metabolite identification using LC coupled to high-resolution MS. *Bioanalysis, 5,* 1285–1297.

Mócsai, A., Kovács, L., & Gergely, P., (2014). What is the future of targeted therapy in rheumatology: Biologics or small molecules? *BMC Med., 12*, 33–43.

Morin, L. P., Mess, J. N., & Garofolo, F., (2013). Large-molecule quantification: Sensitivity and selectivity head-to-head comparison of triple quadrupole with Q-TOF. *Bioanalysis, 5*, 1181–1193.

Nau, R., Sörgel, F., & Eiffert, H., (2010). Penetration of drugs through the blood-cerebrospinal fluid/blood-brain barrier for treatment of central nervous system infections. *Clin. Microbiol. Rev., 23,* 858–883.

Nishizuka, S., Charboneau, L., Young, L., Major, S., Reinhold, W. C., Waltham, M., Kouros-Mehr, H., et al., (2003). Proteomic profiling of the NCI-60 cancer cell lines using new high-density reverse-phase lysate microarrays. *Proc. Natl. Acad. Sci. U S A., 100,* 14229–14234.

Nissim, A., & Chernajovsk, Y., (2008). Historical development of monoclonal antibody therapeutics. In: Chernajovsky, Y., & Nissim, A., (eds.), *Therapeutic Antibodies, Handbook of Experimental Pharmacology* (pp. 3–18). Springer-Verlag: Berlin Heidelberg.

Nwaka, S., & Ridley, R. G., (2003). Virtual drug discovery and development for neglected diseases through public-private partnerships. *Nat. Rev. Drug Discov., 2,* 919–928.

Pagé, M. G., Kudrina, I., Zomahoun, H. T. V., Ziegler, D., Beaulieu, P., Charbonneau, C., Cogan, J., et al., (2018). Relative frequency and risk factors for long-term opioid therapy following surgery and trauma among adults: A systematic review protocol. *Syst. Rev., 7*, 1–11.

Prima, A. A., (2008). *A Study of Using Fuzzy Logic Based System for Adaptation of Precise Insulin Dosage for Type II Diabetes Patients*. Project report, Bachelor of Pharmacy, BRAC University.

Rago, B., & Negahban, A., (2016). Outsource-ability of HRMS-based DMPK assays. *Bioanalysis, 8*, 1641–1644.

Ramanathan, D. M., (2013). Looking beyond the SRM to high-resolution MS paradigm shift for DMPK studies. *Bioanalysis, 5*, 1141–1143.

Ramanathan, R., & Korfmacher, W., (2012). The emergence of high-resolution MS as the premier analytical tool in the pharmaceutical bioanalysis arena. *Bioanalysis, 4,* 467–469.

Ramanathan, R., & Korfmacher, W., (2016). HRMS or HRAMS. *Bioanalysis, 8*, 1639–1640.

Ramanathan, R., Jemal, M., Ramagiri, S., Xia, Y. Q., Humpreys, W. G., Olah, T., & Korfmacher, W. A., (2011). It is time for a paradigm shift in drug discovery bioanalysis: From SRM to HRMS. *J. Mass Spectrom., 46,* 595–601.

Rang, H. P., & Vasella, D., (2006). *Drug Discovery and Development-Technology in Transition.* Elsevier.

Roy, A., Nair, S., Sen, N., Soni, N., & Madhusudhan, M. S., (2017). In silico methods for design of biological therapeutics. *Methods, 131,* 33–65.

Samish, I., MacDermaid, C. M., Perez-Aguilar, J. M., & Saven, J. G., (2011). Theoretical and computational protein design. *Annu. Rev. Phys. Chem., 62,* 129–149.

Scudiero, D. A., Shoemaker, R. H., Paull, K. D., Monks, A., Tierney, S., Nofziger, T. H., Currens, M. J., et al., (1988). Evaluation of a soluble tetrazolium/formazan assay for cell growth and drug sensitivity in culture using human and other tumor cell lines. *Cancer Res., 48,* 4827–4833.

Singh, N. K., (2017). microRNAs databases: Developmental methodologies, structural and functional annotations. *Interdiscip. Sci., 9,* 357–377.

Stolker, A. A. M., Niesing, W., Fuchs, R., Vreeken, R. J., Niessen, W. M. A., & Brinkman, U. A. T., (2004). Liquid chromatography with triple-quadrupole and quadrupole-time-of-flight mass spectrometry for the determination of micro-constituents – a comparison. *Anal. Bioanal. Chem., 378,* 1754–1760.

Sturm, R. M., (2016). High resolution accurate mass spectrometry for quantitative regulated bioanalysis: How, when, and why to implement. *Bioanalysis, 8,* 1709–1721.

Talati, N. J., Rouphael, N., Kuppalli, K., & Franco-Paredes, C., (2008). Spectrum of CNS disease caused by rapidly growing mycobacteria. *Lancet Infect. Dis., 8,* 390–398.

Tanzi, R. E., & Bertram, L., (2001). New frontiers in Alzheimer's disease genetics. *Neuron, 32,* 181–184.

Touw, D. J., Neef, C., Thomson, A. H., & Vinks, A. A., (2005). Cost-effectiveness of therapeutic drug monitoring: A systemic review. *Ther. Drug Monit., 27,* 10–17.

Venter, J. C., Adams, M. D., Myers, E. W., Li, P. W., Mural, R. J., Sutton, G. G., Smith, H. O., et al., (2001). The sequence of the human genome. *Science, 291,* 1304–1351.

Volinia, S., Calin, G. A., Liu, C. G., Ambs, S., Cimmino, A., Petrocca, F., Visone, R., et al., (2006). A microRNA expression signature of human solid tumors defines cancer gene targets. *Proc. Natl. Acad. Sci. U S A., 103,* 2257–2261.

Wallace, Jr. R. J., Musser, J. M., Hull, S. I., Silcox, V. A., Steele, L. C., Forrester, G. D., Labidi, A., & Selander, R. K., (1989). Diversity and sources of rapidly growing mycobacteria associated with infections following cardiac surgery. *J. Infect. Dis., 159,* 708–716.

Wang, Y., & Lee, C. G., (2010). MicroRNA and cancer-focus on apoptosis. *J. Cell Mol. Med., 13,* 12–23.

Wei, H., Tymiak, A. A., & Chen, G., (2013). High-resolution MS for structural characterization of protein therapeutics: Advances and future directions. *Bioanalysis, 5,* 1299–1313.

Wei, J., Gao, W., Zhu, C. J., Liu, Y. Q., Mei, Z., Cheng, T., & Shu, Y. Q., (2011). Identification of plasma microRNA-21 as a biomarker for early detection and chemosensitivity of non-small cell lung cancer. *Chin. J. Cancer, 30,* 407–414.

Weng, N., (2016). A workflow for absolute quantitation of large therapeutic proteins in biological samples at intact level using LC-HRMS. *Bioanalysis, 8,* 1679–1691.

Winthrop, K. L., Abrams, M., Yakrus, M., Schwartz, I., Ely, J.,Gillies, D., & Vugia, D. J., (2002). An outbreak of mycobacterial furunculosis associated with footbaths at a nail salon. *N. Engl. J. Med., 346,* 1366–1371.

Wu, W., Sun, M., Zou, G. M., & Chen, J., (2010). MicroRNA and cancer: Current status and prospective. *Int. J. Cancer, 120,* 953–960.

Xian, F., Hendrickson, C. L., & Marshall, A. G., (2012). High resolution mass spectrometry. *Anal. Chem., 84*, 708–719.

Yang, N., Kaur, S., Volinia, S., Greshock, J., Lassus, H., Hasegawa, K., Liang, S., et al., (2008). MicroRNA microarray identifies Let-7i as a novel biomarker and therapeutic target in human epithelial ovarian cancer. *Cancer Res., 68*, 10307–10313.

Zelesky, V., Schneider, R., Janiszewski, J., Zamora, I., Ferguson, J., & Troutman, M., (2013). Software automation tools for increased throughput metabolic soft-spot identification in early drug discovery. *Bioanalysis, 5*, 1165–1179.

CHAPTER 5

Quality Management in a Feasible Manner Through HRMS

SHINTU JUDE and SREERAJ GOPI

R&D Center, Aurea Biolabs (P) Ltd., Kolenchery, Cochin – 682311, Ernakulam, Kerala, India, E-mails: shintujude2@gmail.com (S. Jude); sreerajgopi@yahoo.com (S. Gopi)

ABSTRACT

A great part of product quality assurance (QA) deals with analyzes. So, the parameters, methods, and analysis setups are to be validated and confirmed. HRMS is a perfect suggestion for many of the analysis needs, due to its peculiar properties such as high resolution (HR), accurate mass measurement, sensitivity, selectivity, retrospective nature of acquired data, etc. Here, in this chapter, we discuss the possibilities and recent studies of HRMS, in the field of quality management.

5.1 INTRODUCTION

The term 'quality management' covers all the acts and approaches to ensure the quality of products or services provided. Related to the products, it is a universal practice to analyze different parameters to ensure the quality of them. The characteristics of quality can vary with the product. It can be the assurance of the presence or absence of particular components, physicochemical characteristics, durability properties, working

High-Resolution Mass Spectrometry and Its Diverse Applications: Cutting-Edge Techniques and Instrumentation. Sreeraj Gopi, PhD, Sabu Thomas, PhD, Augustine Amalraj, PhD, & Shintu Jude, MSc (Editors)

conditions, etc. But doubtlessly, these characteristics are all meant to assure customer satisfaction. Quality management processes generally involve quality assurance (QA) and quality control (QC). As per the description provided by ISO9000, QA is the segment, which addresses the activities of the laboratory to provide confidence that quality requirements will be fulfilled, and QC furnishes the individual measures which are used to actually fulfill the requirements (ISO 9000:2015). Here, in the case of QC of the products, high resolution mass spectrometer (HRMS) can serve some help. Nowadays, many of the products are reaching to the customer hands after labeling 'quality assured,' aiding HRMS for some aspects.

Quality management deals with the assurance of both presences of necessary moieties and absence of impurities. So, the number of analytes to be handled are high, which raises the hurdles such as analysis time, the amount of sample required for the effective methods, mutual interferences of analytes, amount of consumables required for sample preparation and analysis protocols, etc. HRMS provides the way for quality management, by dealing with all these considerations owing to its powers of high resolution (HR), high selectivity, full scan sensitivity, scan speed and mass accuracy (MA) (Gomez et al., 2013). The excellence in resolution provides the possibilities of multi-residue analysis with lesser time. Multiresidue analysis deals with the detection, identification, and quantification of a number of compounds even in trace amounts, without any matrix effects, which forms the major hurdle in natural products analyzes. Mass resolving power and accurate mass measurements contribute towards the narrow mass extraction windows, which in turn achieve high selectivity and can answer the scenario of the lower amount of samples and higher number of compounds. High sensitivity also contributes towards the detection of trace compounds. Even though the concept of untargeted screening (which was assigned to be the major property of HRMS) has not been vanished completely, there are many studies proving that HRMS is capable of providing absolute quantification. On comparing the quantitative performance of HRMS with that of triple quadrupole (QqQ)-MS, it was providing reliable and robust results. Also, the high resolution full scan (HR-FS) runs allowed a relative quantitative measurement of the non-targeted compounds (Grund et al., 2013). Besides, it is important that much of the interferences encountered while using a low resolution mass spectrometer were effectively reduced by HRMS. In case of closely related compound analogs, HRMS offers the proper pick of the molecules of interest by exact mass measurement (Schwarzenberg et al., 2014). This helps in the separation of isotopic peaks of an analyte. The new generation HRMS techniques allow the analysis of samples at a femtomole level, which reduces the hustle of sample availability or its rare occurrence (Neumann et al., 2018). Apart from the conventional quality assessment systems, high MA from HRMS provides a reliable platform for the analysis of unknowns, which ensures a second line of quality management. Besides, the HR-FS runs allow the interpretation of acquired data, in different formats, even after a long time post running the samples (Gomez et al., 2013).

The analytical QA measures have to ensure the confidence in the compound identification, without any false positives (Type I errors). Similarly, false negatives should also be avoided (Type II errors). The primary conditions, making this a challenge is the matrix interferences affecting the analyte of interest and the lower concentrations of the compound. The secondary challenges are (a) the diversity of samples and their matrices to be handled for a single compound and (b) occurrence of the unknown/new compounds (can be a contaminant or necessary compound). These secondary challenges are hard to answer with usual techniques, but with non-targeted analysis protocols proposed by HRMS. The capability of HRMS to deal with both the targeted and untargeted strategies make them a strong platform for a number of analytical interrogations. Table 5.1 gives a basic idea on the differences between targeted and untargeted protocols.

TABLE 5.1 Comparison between Targeted and Untargeted Protocols

Workflow Parameters	Targeted Protocol	Untargeted Protocol
Sample preparation	Elimination of maximum interfering compounds to isolate the analytes of interest	Reduces matrix effect and retain the maximum compounds of interest
Acquisition parameters	Predetermined ions of definite compounds of interest only	A range of masses and fragmentation patterns with accurate measurement
Data analysis	Comparison with external or internal standards for verification and quantification	Detection and identification of compounds and their features; comparison of different samples, even without identification

The chapter depicts some of the important fields, where HRMS is a part of QC measures and Figure 5.1 represents an outline.

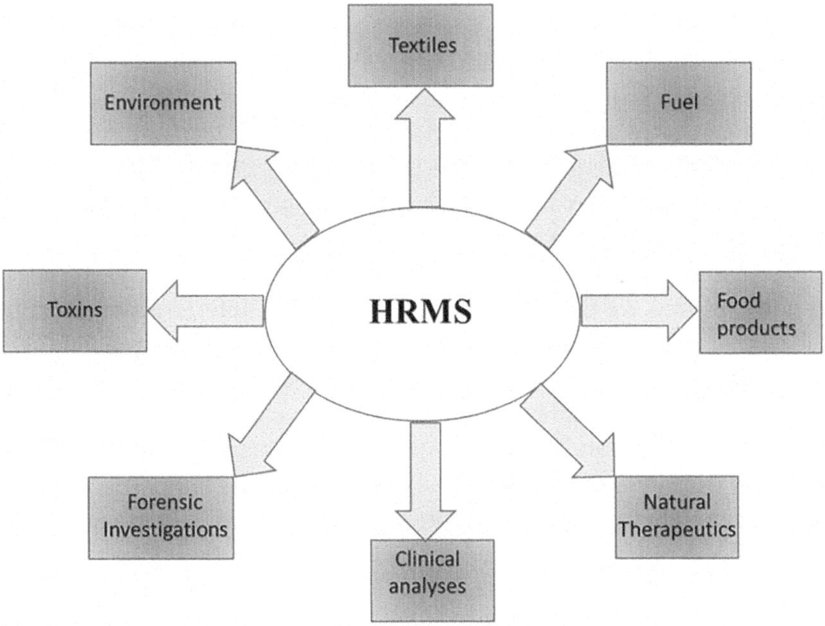

FIGURE 5.1 Different fields, where HRMS serve in the quality control.

5.2 NATURAL PRODUCTS

Analysis of natural products becomes more complicated, primarily due to the complex nature of their matrix. Here, it is really hard to standardize analysis protocols for particular compounds, rather need to consider the sample matrix and conditions. There are many areas/industries hold natural products as the key factor, discussion of which is beyond the scope of a single chapter. So, we can go through two important, closely related streams of natural products – food and natural therapeutics – which are directly related to everyday human life and hence seek more QA parameters.

5.2.1 FOOD

In the food industry, the quality parameters include assurance on safety profiles, residual impurities, geographical origin, organoleptic properties, presence of characteristic components and contamination related to microbes, additives, allergens, or adulteration. From an industrial point of view, where the food products need to get standardized, compliance issues arise from raw material species variations, agro-climatic impacts, samples authentication, etc. HRMS finds place as part of the many analysis systems to find solutions for these problems. In case of QA or QC of food, there are many validation guidelines are available from authorities, including The United States Food and Drug Administration (USFDA), European Pharmacopoeia, US Pharmacopoeia, Association of Official Analytical Chemists (AOAC), International Conference on Harmonization, Food Safety Standard Authority of India, etc. (Fibigr et al., 2018). Not only for food, but for all the quantitative assessments (of both bioactive compounds and contaminants), the methodology needs to get validated with parameters such as system suitability, selectivity, linearity, precision, accuracy, robustness, recovery, interferences, LOD, LOQ, and stability of analyte in the sample. Different food quality parameters are represented in Figure 5.2.

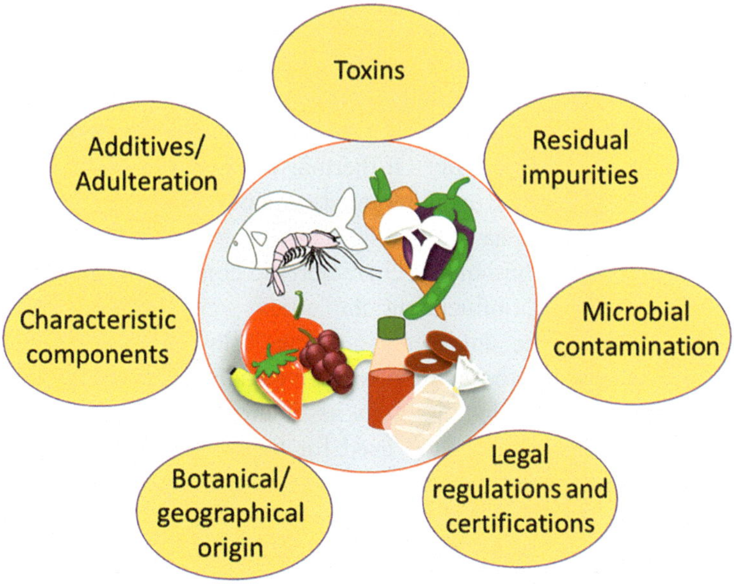

FIGURE 5.2 Different quality control parameters interfering with food materials.

It is assessed that almost 90% of the food products are agricultural products (vegetable products). The nutritional profile (including the amount of vitamins, proteins, minerals, fats, and carbohydrates), flavor parameters (volatiles, sugars, astringency, and acidity) and even the moisture content are susceptible to vary with the environmental and geographic conditions as well as agricultural practices (Salunkhe et al., 1988). Besides the agricultural practices and climate, one of the major factors affecting the quality and value of agricultural products is post-harvest handling and storage procedures. The nature of quality loss occurs during storage depends on the products. For example, in the case of table grapes, it appears as the rachis and berry browning, loss of turgor of berries, weight loss, and even microbial spoilage (Innamorato et al., 2020). While considering milk and milk products, the storage is threatened by microorganisms, resulting in curdling. For coconut water, the deterioration is associated with color change, which can appear with sterilization also (Cunha et al., 2020). These kinds of quality issues were traditionally rectified by sensory and gustatometric evaluations. But a more specific and reliable set of data can be provided by instrumental analyzes, which is possible because any of the metabolic changes are related to changes in the presence and amount of chemical components. By identifying the compounds which contribute towards the quality parameters, it becomes easy to characterize and discriminate the product between different quality levels. In a study, three non-targeted analysis methods viz. HS-SPME, NMR, and LC-HRMS were brilliantly connected for the quality assessment of table grapes (Innamorato et al., 2020). In a similar way the secondary metabolites, acting behind the color change of coconut water on thermal treatment, were profiled with UPLC-HRMS. Here also, the untargeted strategies were used for chemical markers' selection (Cunha et al., 2020).

The presence and amount of characteristic compounds are one of the major considerations, influencing the quality, genuineness, and value of food products. The untargeted analyzes allow the investigation of unknowns, which, together with quantitative analysis abilities, furnishes a perfect platform for food authenticity assessments. The researches which include HRMS were effectively used for the components identification and verification. A study conducted in 2014 successfully used Orbitrap-Exactive-HCD in order to quantify 16 volatile thiols and identify 19 non-targeted thiols from coffee matrix (Vichi et al., 2014). Marchal et al. (2016) developed an effective method using Exactive Orbitrap mass spectrometer with a heated electrospray ionization (HESI) for the quantification of the taste active triterpenoids from extracts of two different species of oak–*Quercus petraea* Liebl. (Sessile oak) and *Quercus robur* L. (pedunculate oak). A triterpenoids index (TI) was calculated using the concentrations of both the sweet triterpenoids-quercotriterpenosides (QTTs) I, II, and III, and bitter triterpene – a glucosyl derivative of bartogenic acid (Glu-BA). The TI value gives an idea about the chemical composition of oak wood and can be used to identify the oak species and its organoleptic effect on wine and spirits (Marchal et al., 2016). A validated short timed method for simultaneous quantitative determination of 21 nutrients from rice, by using UHPLC-LTQ-Orbitrap HRMS have been reported by Li and coworkers (Zhu et al., 2016). Using a DART-MS fingerprinting, Kim et al. designed a method for the identification and QC of cubeb berries (*Piper cubeba* and *Litsea cubeba*) (Kim et al., 2011). With the help of direct analysis in real-time coupled to HR mass spectrometry (DART-HRMS), a set of 25 biomarkers was created, and a comparative quality assessment is made for three milk samples derived from different diets (Riuzzi et al., 2021). Hrbek and the team demonstrated the effective utilization of a single HRMS platform (DART) for several scenarios related to milk authentication. They could discriminate milk from different animal species, differentiate cow's milk produced from organic and conventional farming and could detect the vegetable oil adulteration in soft cheese (Hrbek et al., 2014).

The identification of toxins and adulterants present in the food products is really important. Sometimes the threat is unidentifiable, especially when the toxins are present naturally. Such contaminants can include heavy metals, pesticide residues, microbial toxins, illegal dyes, etc. For example, the alkenylbenzenes present naturally in condiments are endogenous compounds, which are found to be genotoxic and carcinogenic (Alajlouni et al., 2016). Accurate mass measurement, high selectivity and sensitivity of orbitrap was used intelligently for the determination of eight different alkenylbenzenes from the frequently consumed pepper varieties, providing a deep knowledge on the effects of maturity and geographical variations on their contents (Rivera-Pérez et al., 2020). This method provides an option for both routine analysis and a track of consumer exposure towards alkenylbenzenes. A similar, promising quantitative tool for the xenobiotic residues was developed with UHPLC-Q-Orbitrap, which could successfully determine 157 compounds from honey, within a significantly short time of 15 minutes (Li et al., 2016). Another important adulteration with dyes could be quantitatively monitored even at very low concentrations in different matrices, with the help of HRMS (Dubreil et al., 2019; Périat et al., 2019). A quantitative LC-HRMS/MS method, which could detect the presence of four allergens of different origins was developed by Stella et al. They have checked the suitability of the method in commercially available food products, even with high protein content (Stella et al., 2020).

In the case of animal derived products, the residues of veterinary drugs or animal husbandry chemicals are considered as impermissible. Contaminated feeds of animals also happened to be a threat in the meat products (Whitehead, 1998). For example, meldonium, a suspected contamination from veterinary drugs and strictly prohibited under WADA norms, was quantified from milk and poultry meat (Temerdashev et al., 2020). Similarly, the possibility for the presence of antimicrobial and veterinary drugs from distillers grain (by-product of ethanol distillation, which is considered as a valuable feed material for livestock) was evaluated with LC-HRMS and the method was recommended for the QC measures of feed safety (Kaklamanos et al., 2013). There are effective HRMS methods available for the determination of pesticide residues such as cyromazine, melamine, etc., from milk and milk products (Chen et al., 2015).

Microbial contaminations in food are related to the presence or activity of microorganisms. Likewise, the presence of mycotoxins (microbial toxins) produce harmful effects on even healthy bodies. A multi-class toxins analysis methodology has been developed with exactive orbitrap, which was applied for 10 green tea samples and eight royal jelly samples. A total of nine samples was found to be positive towards pesticides and mycotoxins with an LOQ of <10 μg/kg (Martínez-Domínguez et al., 2016). Similarly, a simultaneous full-scan and fragmentation data acquisition was performed for 184 pesticide residues present in two different matrices by utilizing LC-HRMS. The methodology put forward the option of a single acquisition for both screening and quantification of compounds (Zomer, 2015). Rather, HRMS is a reliable platform for the analysis of a wide range of pesticide residues, from different matrices (Gómez-Ramos et al., 2013; Pico et al., 2020).

Possibilities of untargeted investigations can lead the capabilities of HRMS towards unexpected inventions as resulted with Bengtström et al. (2016). They have identified the contaminants such as dioxins from recycled paper and board food-contact materials. With the help of QTOF MS, they have tentatively identified 75 compounds. The questions on the safety of recycled food packaging compounds, which they have raised was seconded by the 3% aryl hydrocarbon receptor activity exhibited by the suspected fractions (*in vitro*) (Bengtstrom et al., 2016).

Metabolomics is an important stream in science with uncountable outcomes. Food metabolomics, based on MS is a perfect platform for multi-component determination, especially for contaminants such as pesticides, mycotoxins, veterinary drugs, etc., which may influence the formation of metabolites (Castro-Puyana et al., 2017). Interestingly, differentiation between two grape cultivars made successful by employing metabolites profiling of the same with GC-TOF-MS, which in turn utilized for the profiling of metabolic pathways and their relations to the growth. This contributes towards the QC, as the amount of metabolites determines the growth and vice versa (Cuadros-Inostroza et al., 2016). Another important innovation in this field is the determination of 'microbial volatile organic compounds' (MVOC) as the microbial contamination markers. As the name indicates, specific volatile compounds are determined in correlation with the growth and activity of particular pathogens, which is then used for multivariate analysis. There are success stories reported for this methodology, in evaluating the chemical spoilage and prediction of shelf-life (Castro-Puyana et al., 2017). Metabolites profiling was logged for the determination of the places of origin, conditions of processing and even the ripening period of 12 kinds of dry-cured ham, which extends the applications towards traceability (Sugimoto et al., 2017). Rather, the metabolite profiling is an effective model for crossing the hurdles of matrix effects. There are reported successful food metabolomics protocols available for different matrices such as baked foods, fruits, vegetables, honey, nuts, tea, seafoods, etc. (López-Ruiz et al., 2019).

5.2.2 NATURAL THERAPEUTICS

The value-added upgrades of food, known as dietary supplements are important products of the new era. The dietary supplements, known sometimes as nutraceuticals act for complementing the food. They serve a step between food and medicine. Most of them are valued for their bioactive components and hence, activity assessment and component determination are different modes of QA, implied for bioactive components. Besides, nutraceuticals, especially of natural origin find great popularity, which generates competitions in market and of course face the threat of adulteration, contaminations, safety, and quality issues. So, the determination of adulterants also comes into the picture. Quantitative multicomponent analyzes require more number of standards, lengthy analysis time, different processing protocols, etc. The possibilities of HRMS were used well in many studies to reduce these snags. For example, by using only a single marker compound, 13 bioactive components were able to quantify from walnut leaves by UHPLC-Q-Orbitrap HRMS. The efficiency and accuracy of the proposed method, (commonly known as 'quantitative analysis of multi-components by a single marker (QAMS)') were evaluated by an auxiliary method with all the 13 reference standards. Exact mass determination, HR and sensitivity allow the determination of different components without reference standards (Su et al., 2020). UPLC-QqQLIT (triple quadrupole linear ion trap) instrumental setup was used in another study, for the successful development and validation of a method for the simultaneous determination of 26 multiclass bio-active components from *Garcinia* leaf extract. Around 11 different species were verified with this method, establishing the fact that the remarkable chemical difference is there between each species. This method-validated under the norms of the International Conference on Harmonization – is a reliable tool for the quality determination, authenticity assessment and selection of proper raw material (Pandey et al., 2015). Chandra et al. (2015) developed a method for the quantitative determination of 18 compounds in different *Cassia* species, using the instrumentation setup UPLC-QqQLIT. The method was validated and used for the investigations on the variations of bio-actives among different plant parts in five *Cassia* species. A supplementary study for in-vitro anti-proliferative effects of their methanol extracts, using sulforhodamine B (SRB) assay on five different cell lines found >50% growth inhibition (Chandra et al., 2015). The cannabinoids fingerprinting of macerated oil of two authorized medical-grade varieties of *Cannabis sativa* L. was successfully achieved by the targeted and untargeted profiling using HPLC-HRMS hyphenation (Calvi et al., 2018). In a similar way, quantification of the phytochemicals presented in 11 different nutraceuticals derived from *Ginkgo biloba* has been carried out using UHPLC-Orbitrap-MS. There was a huge difference in the terpenoids content (from 1,133 mg/kg to 12,706 mg/kg), even within the 11 samples, which raises the need for compulsory quality verification of *Ginkgo* products (López-Gutiérrez et al., 2016).

In many cases, quality assessment is associated with activity validation. The therapeutic efficacy of an edible herb named Argel, traditionally used for rheumatoid arthritis was found to attribute to its secondary metabolites. Likewise, a coupling of the DPPH scavenging assay with UHPLC-HRMS enabled the detection of true antioxidants and active compounds present in black tea. The profiling included detection of antioxidants in trace amounts, resolution of co-eluted antioxidants and structural characterization of unknowns (Chen et al., 2020). Sometimes, natural therapeutics contain two or more materials/ingredients, which make the profiling and quality evaluation a great task. For example, *Fufang Xueshuantong Capsule* (FXC), a traditional Chinese medicine (TCM) containing four different herbs was subjected to a number of studies to establish quality and efficacy evaluation schemes. According to the Chinese Pharmacopoeia, the capsule's quality evaluation includes the assessment of medicinal materials, adulteration with other organs of the same plant than required and quantitative determination of active compounds (Zeng et al., 2020). Together, it was not easy to handle all the parameters at once until HRMS is there in the picture. By using UHPLC-IM-QTOF, a comprehensive multi-component characterization was demonstrated for 230 components from FXC (Zuo et al., 2019).

It is important in QA systems to verify the maximum components as fast as possible, with accurate measurements. One such goal was achieved by Goessens and crew, where they have analyzed 46 antimicrobial drug residues belonging to different classes, from pond water, using UHPLC-Orbitrap-HRMS (Goessens et al., 2020). The method becomes more important, with its shorter elution time of <18 minutes with proper separation of all the peaks. Likewise, 4 minutes was enough for the quantification of 11 systemic antimycotics by using orbitrap-HRMS (Schuster et al., 2019). In the case of dietary supplements, the possibilities of adulteration with intentionally added pharmaceuticals can cause many health problems. Also, products of natural origin face the threats from mycotoxins and pesticides residues, etc., as discussed earlier. QA in terms of adulterants can be achieved in a feasible manner with HRMS. For example, a multicomponent analysis platform was furnished with exactive orbitrap, which possesses the capability for the identification of more than 250 toxic substances, belonging to different classes such as mycotoxins and pesticides. Real sample analysis was demonstrated for the method, with 3 different dietary supplements of soy origin, out of which, four samples have produced positive results (Martínez-Domínguez et al., 2016).

5.2.3 ORIGIN DETERMINATION

Origin determination is really important for agricultural products as the quality of the same is related to their origin, in many cases. Origin determination come into action for food quality management, forensic applications, scientific evidences for illegal cultivations and smuggling, etc. Plant derived products are always in the shade of unidentifiable adulteration threat, as the final product results from a number of processing steps such as pulverization, extraction, temperature treatment, blending, chemical treatment, etc., which can destroy the visual indicators. For some natural products, the production area/geographical origin is a quality determining factor and the same has been subjected to fraudulence for value. Similarly, illegal harvesting and trading of endangered plants are not easily provable, as there are not much efficient conventional methods to identify the origin of species. Concerning all these facts, many methods, including DNA analysis, infrared spectroscopy (IR) and NMR are proposed to identify and differentiate the botanical origin of natural products (Raclariu et al., 2017, 2018; Minaei et al., 2017; Giraudo et al., 2019; Zhang et al., 2018). But all of these methods have met with either time consuming analysis procedures, consumption of costly resources, or tedious extraction steps, where the biomarkers may not be withdrawn completely.

DART-HRMS is an effective solution for the fore-stated drawbacks. It does not require any sample processing and can analyze the sample in its native form. So the issues such as sample polarity, dissolution properties, expensive consumables, etc., will not arise. Also, samples of different physical forms can be analyzed using a single method. Many studies have proved the efficiency of DART-HRMS for the identification of species by deriving a chemical signature profile. This instrument derived species biomarker identification has been used successfully to prove that, *Eucalypts* of different species exhibit production of temperature dependent volatile oil with unique nature (Maleknia et al., 2009). The same system was proved useful to differentiate between two different basil cultivars and eight betel (*Piper betle* Linn.) cultivars (JEOL, 2006; Bajpai et al., 2010). The origin of the cubeb berries from different species named *Lauraceae* and *Piperaceae* were identified with DART-MS system (Kim et al., 2011). Provenance determination of the psychoactive pepper species, namely *Piper methysticum* (aka kava) and *P. betle* (aka betel) were carried out by using DART-HRMS, by principal component analysis (PCA) of the samples in their native form, without much sample preparation steps (Lesiak and Musah, 2016).

A study published in 2015 presented a consisted and reproducible single method in which, by maintaining the same conditions, a range of chemotypes from different samples were detected in DART coupled to a TOF-MS for their species classification. The method demonstrated the uses thereof, such as identification of endangered *Dalbergia granadillo Pittier* from the other species commonly known as granadillo, determination of FAME (fatty acid methyl ester) profile of biodiesel feedstock for source identification, creation of the hydrocarbon profile of the puparial cases of different blowflies to be used in forensic investigations, determination of eucalyptus species by analyzing the leaf, and finally, identification, and differentiation of the different species of the genus Datura from other psychoactive plants, which also produce the biomarker compound of the former–belladonna alkaloid by analyzing the seeds (Musah et al., 2015). DART-TOFMS was successfully used to discriminate geographic provenances of plants from the same species. Studies have been conducted to differentiate the specimens such as *Angelica gigas* roots from China and Korea (Kim et al., 2015), cultivated, and wild species of *Aquilaria Lam.* spp. (Espinoza et al., 2014) and *Douglas-fir* (*Pseudotsuga menziesii* (Mirb.) *Franco var. menziesii*) from the Oregon Coast Range and Oregon Cascade Range (Finch et al., 2017). A combination of HPLC-PTR (proton transfer reaction) – TOF MS-PCA (principal component analysis) was successfully employed for the determination of geographical origin of saffron samples, choosing safranal, picrocrocin, and isophorone as the informative compounds (Masi et al., 2016).

Origin determination is an important segment in adulteration finding and authentication of products and is important for the industries which tag the product price according to the provenance. For example, the varying cost of Iranian saffron, Arabica coffee, olive oil, cheese, wine, etc., is depending on their geographical origin. Many studies involved HRMS in different formats in the instrument team for the adulteration testing and origin determination of natural products (Cavanna et al., 2018). A hyphenation of HPLC-CZE-HRMS combined with PCA data treatment for finding adulteration in cranberry extracts, HPLC-Q-TOF HRMS for Chinese traditional patent medicines (CTPM), direct infusion in HRMS for milk, orbitrap MS for herbal medicines and dietary supplements, FT-IR HRMS for herbs, UHPLC-HRMS for green tea, LC-HRMS with metabolomics data for coffee, etc., are some examples of this kind (Navarro et al., 2014; Wang et al., 2018; Guerreiro et al., 2018; Shi et al., 2014; Black et al., 2016; Souard et al., 2018). Even a more complex matrix like honey was well discriminated for their geographical origin, by using the untargeted DART-HRMS analysis, in combination with chemometrics (Lippolis et al., 2020).

Information about the origin, sometimes plays a major role in archaeology and history also. Old day dye sources are not easy to identify with usual analysis methods. But recently some successful stories have made on the same. For example, by using UHPLC-Q-Tof MS system, *Sophora japonica* L. and *Phellodendron amurense* Rupr. were identified as the dye sources for the yellow color of ancient Chinese textiles from the Qing Dynasty Royal Palace. A color restoration experiment was conducted to confirm the results. The analysis data provided light towards the stable natural color pigments as well as the dyeing technology of ancient China (Zhang et al., 2017).

5.3 CLINICAL AND RELATED FORENSIC FIELD

HRMS is widely used in the clinical innovations and therapeutic drugs (TDs) as well as vaccine designing. The instrumentation can contribute towards the product characterization and activity identification, which in turn play a role in product quality. Yu et al. made use of a synergistic combination of HRMS and NMR for the analytical characterization of Glycoconjugate Vaccines, which was considered as tedious due to the heterogeneous nature of their sites of glycosylation as well as oligosaccharide length. The improved characterization protocol has ended in a better understanding of the mechanism of conjugation (Yu et al., 2018). It is really important to assure the safety of products, along with their efficacy, especially in the pharmaceutical industry. In a study, a workflow is constructed using UHPLC-HRMS series to profile the impurities that may present in the synthetic thyroid hormone (thyroxine). A total of three minutes was enough to knock out 71 impurities from the samples

(Neu et al., 2013). Along with the rapid multicompound detection, the method raises platform for the semiquantitative determination of even trace amount of components. Similarly, a peptide drug product quality monitoring approach was developed by using LC-HRMS, by which, it was possible to determine the amino acid composition, confirm the peptide sequence, and quantify any peptide related impurities, even when they are co-eluting, simultaneously within a single experiment (Zeng et al., 2015).

The analysis of pharmaceutical samples of any form is made possible with the advancements in HRMS. By using desorption electrospray ionization (DESI), the components (including impurities) even present in trace amounts were identified from therapeutic formulations in the form of tablets, ointments, and liquids that too without prior sample preparation. DESI enables a high sampling rate in an ambient environment, without carryover effects. The properties altogether furnish a platform for the online high throughput screening of pharmaceutical and alike samples (Chen et al., 2005). In another study, a QC platform for the release and stability testing of a commercial antibody product was developed and was implemented to ensure the quality, in terms of post-translational modifications (PTMs) critical quality attributes (Sokolowska et al., 2019).

Structural knowledge on the medications and targets are important in the therapeutic field. As HRMS provides its best for structure elucidations, it holds the priorities in the field. Solouki et al. (1997) demonstrated different strategies, utilizing MALDI FTICRMS for the determination of the number of sulfur atoms and disulfide bridges in biologically active peptides (Solouki et al., 1997). A similar work was presented by Schnaible and team for screening for disulfide bonds in proteins by using MALDI in-source decay and LIFT-TOF/TOF-MS (Schnaible et al., 2002). Similarly, the elemental composition determinations for larger molecules such as DNA, RNA, and oligonucleotides help for the base composition assessment, which in turn allow their sequence determination (Hofstadler et al., 2005; Orr et al., 2019). Another important application of HRMS in this field is polymerase chain reaction (PCR) products analysis (Hahner, 2002; Krahmer et al., 1999). These findings allow the analyzes of macromolecules for forensic applications also.

HRMS hyphenations found useful in the functional characterization of organs. The multiplex detection and characterization of a number of compounds from a single sample, detecting even at femtomole concentration is a versatile property exhibited by matrix-assisted laser desorption/

ionization-mass spectrometry imaging (MALDI-MSI) hyphenation. The MSI-IR sequential analysis was used for the spatio-chemical characterization of different biological samples in a study. The data set was subjected to chemical pan sharpening, which provides the ion distribution images. This kind of data fusion enabled the distinction of relevant brain regions (Neumann et al., 2018).

In therapeutic systems also, natural products play a major role. Unlike the synthetic drugs, it is difficult to produce standardized natural products with sustained composition. Hence protocols are needed to develop for the standardization of the same. Here also, HRMS helps. For example, *Yinchenhao Decoction* (YCHD), a classical TCM formulation composed of *Artemisia capillaries, Fructus Gardeniae,* and *Rhizoma et Radix Rhei* was found to have a complex matrix. A method for the speedy, accurate detection and structural characterization of this formulation was developed by using HPLC-QToF-MS/MS, which is fit to be used as a QC aid. Find by formula (FBF) algorithms was applied for the screening of YCHD, and not surprisingly, 77 major compounds from the formulation was characterized (Fu et al., 2015). Likewise, a series of analysis methods were derived for the bioactive components of the medicinal mushroom *Antrodia cinnamomea*, making use of UHPLC-DAD-QToF-MS, UPLC/UV, supercritical fluid chromatography (SFC)-MS and ion chromatography coupled with pulsed amperometric detection (IC-PAD). A QC device for the vital triterpenoids ergostane and lanostane was prepared and verified by applying the method for 15 batches of the species (Qiao et al., 2015). Yao et al. (2016) developed a quantitative strategy for a Chinese patent medicine – Shuxiong Tablet (SXT) by establishing a UPLC-QDa-selective ion monitoring (SIM) method, derived from systematic multi-component (SMC) characterization carried out by UPLC-QToF-Fast DDA. The SMC-SIM strategy has been proved to be a potent QC tool and 12 different SXT batch samples were successfully analyzed in this method (Yao et al., 2016).

5.4 CHEMICALS/TOXINS AND ENVIRONMENTAL FACTORS

Chemicals are really a kind of unpredictable species. Only trials and experiments can assure their safety and usage profiles. Some of them are produced with a purpose, some are by-products and some occur or evolve in the nature itself. The identification of unknown compounds and their properties is important in quality point of view. When it comes to environmental samples, the range of unknowns is enormous, which could affect the daily life of many. Water, air, and soil quality determine the health parameters of a community itself. Every industrial processes and product are happened to end up in any of the environmental denomination such as soil, air or water. Many of the chemical moieties produced for use in a positive way, were found to cause several unwanted side effects. For example, chloromethyl ether is an important moiety for the production of anion exchange resins, several aromatic products and membranes. But, one of its analog named bis-chloromethyl ether (BCME) – is found to be extremely toxic. In an earlier study, a highly selective method was developed for the determination of BCME in ppb levels, using GC-MS (Müller et al., 1981). In a similar way, chlorinated paraffins (CPs) are useful in metal and plastic industry, but are recognized as endocrine disruptors and carcinogens. Their residues can be found in a number of matrices, including soil, water, air, meat, human blood and even in breast milk. The major difficulty faced during their analyzes is their large number of congeners and isomers, which in turn cause interferences. But chromatographic separations coupled with HRMS reduce these complications and the high MA as well as resolution provides a strong platform for the proper analyzes of CPs (Huang et al., 2019). Similarly, brominated flame retardants (BFRs) were found to cause adverse effects on the Eco-cycle. Analytical determination of BFRs is not so easy, as they possess distinct physicochemical properties, faces analytical interferences from their own degradation products and the instability of congeners. Methods, equipped with HRMS provide a better platform for BFRs analysis from a range of matrices, such as air, marine samples, cow milk, human milk, fly ash, etc. (Jessika et al., 2009). Polybrominated diphenyl ethers (PBDEs) forms a major group of BFRs and their environmental and health hazards seek attention for advanced quality measures. A picture of atmospheric pollution caused by PBDEs is depicted by Peng et al. using HRGC-HRMS. They have conducted a spatiotemporal variability study of PBDE levels and established the correlations of traffic, industrial, residential, and electronic discharges as well as season towards PBDE levels, which in turn is suggested to prepare the quality monitoring protocols (Peng et al., 2018).

Organic pollutants such as polychlorinated dibenzo-p-dioxins (PCDDs), dibenzofurans (PCDFs), dioxin-like polychlorinated biphenyls (PCBs), brominated dioxins and furans (PBDD/PBDFs) are found to be highly toxic, especially if happen to expose at early stages of life, which result in developmental toxicity. Many studies have suggested the fact that their presence is increasing in the surroundings by any means, which increases the potential health threats (Zhan et al., 2019; van den Berg et al., 2013). One fear factor is that most of the recycled plastic products, including the toys for young children contain these dioxins like contaminants and can cause harm to the children. A proof on the issue was provided by Budin et al. (2020) by assessing the presence and related toxicity of PBDD/Fs from plastic toys, using a hyphenation of GCHRMS and cell-based AhR reporter gene assays in both rats and human genes. The bioanalytical equivalents determination further evaluated the total dioxin intake of children from the toys, which put forward an effective means for QA (Budin et al., 2020).

Water is considered as universal solvent and usually contains metals, organic matters and chemical compounds which can occur naturally or from anthropic interferences (Jessika et al., 2009). The presence of these matters can improve or decrease the water quality, depending on the nature and amount present. So, regular screening of water for its quality is an important practice. There are many monitoring programs and corrective actions are erected by authorities in many countries against such contaminations. One of the possible and practical remedies adapted for the regularly appearing unknown contaminants in the surface water from the river Meuse in The Netherlands was to keep an eye on the same and proceed with the instrument combination of the LC-HRMS-NMR. The detection, identification, and structural confirmation of the unknown compounds was successfully carried out and was verified with pure reference compound (van Leerdam et al., 2014). A similar attempt has been made with the nontargeted screening by HRMS, for screening the organic pollutants on the river Rhine in central Europe. Here, the monitoring made use of three phases of HRMS analysis – quantitative screening of targeted compounds, screening of suspected compounds for peak events identification and nontargeted screening for undetected or unanticipated compounds (Hollender et al., 2017). The study proposes the transition of nontargeted screening towards solution-based applications. The unknowns' identification pave way for the toxicology studies, safety profiling and corrective actions. The taste and quality of drinking water are greatly affected by the organic matters present in it. The efficiency of HRMS for water analysis is made to utilize for the quality assessment of mineral water and tap water from different countries. UPLC-HRMS screening followed by principal components analysis (PCA) evaluation have demonstrated the screening of 47 different organic compounds, without any sample preparation (Ribeiro et al., 2017).

5.5 TEXTILE

Textile industry performs as a necessary part of everyday human life, as it fulfills one of the basic needs. But, a number of chemicals used for the textile production as additives, detergents, coolants, dyes, wetting agents and bleach, as well as the by-products and effluents produced during the process are found to cause adverse effects. There are studies published on the health risks associated with the direct exposure of these toxic chemicals such as flame retardants, aromatic amines, quinoline, bisphenols, benzothiazoles/benzotriazoles, phthalates, aldehydes, and metal particles. Along with the direct human dermal exposure as clothes, they can cause other severe health problems by affecting environmental and food matrices (Rovira and Domingo, 2019). Alkylphenol polyethoxylated (APEOs) and alcohol polyethoxylated (AEOs) are the major non-ionic surfactants widely used in textile industries and are found to cause many disorders and diseases by the direct exposure or through their metabolites (Jahnke et al., 2004). By using an orbitrap, a simultaneous quantitation of 40 APEOs and screening of 160 AEOs was made possible, a compound database was created thereof, which was demonstrated for 40 textile samples (Luo et al., 2017). As we discussed earlier, HRGC-HRMS is proved to be one of the best solutions for BFRs and PBDEs determination. Their use as additives or reactants in textiles and polymers is one major cause of their increased exposure levels. An analysis method based on HRGC-HRMS was successfully employed for PBDE congeners even at low concentrations from textile samples (Shin and Baek). Extend of environmental pollution caused by PBDE was evidenced from the detection of their presence in edible bivalves from the Italian coast of the Adriatic Sea, by using the same technique (Pizzini et al., 2015).

5.6 FUEL

In order to meet the increasing energy needs of different considerations, new energy sources are introduced nowadays. The requirement for clean, sustainable, and renewable energy sources leaves a space for QA protocols.

As a part, the new fuels need to get profiled for their components. Biomass waste is considered as a potential renewable energy source and there are researches ongoing for the conversion of the same into valuable energy forms. By using a 2D GC-HRMS, a detailed chemical characterization of the pyrolysis bio-oils from two biomass waste products, the soursop (*Annona muricata* L.) seed cake and bocaiuva (*Acrocomia aculeata*) seed cake (BSC) were carried out by Nunes et al. (2020). The same system was used for the chemical speciation of different fractions from solvent deasphalting of petroleum and bio-oil co-processing products. It was helpful in the evaluation of the process as well as the products (Pontes et al., 2020).

5.7 CONCLUSION

Quality is a feature which comes across in many points of our day-to-day life. Rather than a personal need, QA is a social commitment. Analysis and verification of different parameters are the major considerations while dealing with the quality of many products. From a wide range of analytical platforms, mass spectrometers, especially, HRMS instruments provide a spectrum of QC applications, due to its specialty features. HRMS alone can act in many beneficial ways. Still, there are limitations and drawbacks for the current setups. Yet, the possibilities of hyphenation are enormous. The technology combinations provide synergistic effects on different QA protocols. HRMS, with its quantitative, qualitative, and simultaneous Quan/Qual modes furnishes a rapid, convenient, and accurate way for QC.

KEYWORDS

- **high-resolution mass spectrometer**
- **matrix effects**
- **multiresidue analysis**
- **quality assurance**
- **quality control**

REFERENCES

Alajlouni, A. M., Isnaeni, F. N., Wesseling, S., Vervoort, J., & Rietjens, I. M., (2016). Level of alkylbenzenes in parsley and dill based teas and associated risk assessment using the margin of exposure approach. *J. Agric. Food Chem., 64*, 8640–8646.

Bajpai, V., Sharma, D., Kumar, B., & Madhusudanan, K. P., (2010). Profiling of Piper betle Linn. cultivars by direct analysis in real time mass spectrometric technique. *Biomed Chromatogr., 24*, 1283–1286.

Bengtström, L., Rosenmai, A. K., Trier, X., Jensen, L. K., Granby, K., Vinggaard, A. M., Driffield, M., & Højslev, P. J., (2016). Non-targeted screening for contaminants in paper and board food contact materials using effect-directed analysis and accurate mass spectrometry. *Food Addit. Contam. Part A, 33*, 1080–1093.

Black, C., Haughey, S. A., Chevallier, O. P., Galvin-King, P., & Elliott, C. T., (2016). A comprehensive strategy to detect the fraudulent adulteration of herbs: The oregano approach. *Food Chem., 1*, 551–557.

Budin, C., Petrlik, J., Strakova, J., Hamm, S., Beeler, B., Behnisch, P., Besselink, H., Van, D. B. B., & Brouwer, A., (2020). Detection of high PBDD/Fs levels and dioxin-like activity in toys using a combination of GC-HRMS, rat-based and human-based DR CALUX® reporter gene assays. *Chemosphere, 251*, 126579.

Calvi, L., Pentimalli, D., Panseri, S., Giupponi, L., Gelmini, F., Beretta, G., Vitali, D., et al., (2018). Comprehensive quality evaluation of medical *Cannabis sativa* L. inflorescence and macerated oils based on HS-SPME coupled to GC–MS and LC-HRMS (q-exactive orbitrap®) approach. *J. Pharm. Biomed. Anal., 150*, 208–219.

Castro-Puyana, M., Pérez-Míguez, R., Montero, L., & Herrero, M., (2017). Reprint of: Application of mass spectrometry-based metabolomics approaches for food safety, quality and traceability. *TrAC Trends Anal. Chem., 96*, 62–78.

Cavanna, D., Righetti, L., Elliott, C., & Suman, M., (2018). The scientific challenges in moving from targeted to non-targeted mass spectrometric methods for food fraud analysis: A proposed validation workflow to bring about a harmonized approach. *Trends Food Sci. Technol., 80*, 223–241.

Chandra, P., Pandey, R., Kumar, B., Srivastva, M., Pandey, P., Sarkar, J., & Singh, B. P., (2015). Quantification of multianalyte by UPLC–QqQLIT–MS/MS and *in-vitro* anti-proliferative screening in *Cassia* species. *Ind Crops Prod., 76*, 1133–1141.

Chen, D., Zhao, Y., Miao, H., & Wu, Y., (2015). A novel dispersive micro solid-phase extraction using PCX as the sorbent for the determination of melamine and cyromazine in milk and milk powder by UHPLC-HRMS/MS. *Talanta, 134*, 144–152.

Chen, H., Talaty, N. N., Takáts, Z., & Cooks, R. G., (2005). Desorption electrospray ionization mass spectrometry for high-throughput analysis of pharmaceutical samples in the ambient environment. *Anal. Chem., 77*, 6915–6927.

Chen, N., Han, B., Fan, X., Cai, F., Ren, F., Xu, M., Zhong, J., et al., (2020). Uncovering the antioxidant characteristics of black tea by coupling *in vitro* free radical scavenging assay with UHPLC–HRMS analysis. *J. Chromatogr. B, 1145*, 122092.

Cuadros-Inostroza, A., Ruíz-Lara, S., González, E., Eckardt, A., Willmitzer, L., & Pena-Cortés, H., (2016). GC–MS metabolic profiling of cabernet sauvignon and merlot cultivars during grapevine berry development and network analysis reveals a stage-and cultivar-dependent connectivity of primary metabolites. *Metabolomics, 12*, 39.

Cunha, A. G., Alves, F. E. G., Silva, L. M. A., Ribeiro, P. R. V., Rodrigues, T. H. S., De Brito, E. S., & De Miranda, M. R. A., (2020). Chemical composition of thermally processed coconut water evaluated by GC–MS, UPLC-HRMS, and NMR. *Food Chem., 324*, 126874.

Dubreil, E., Mompelat, S., Kromer, V., Guitton, Y., Danion, M., Morin, T., Hurtaud-Pessel, D., & Verdon, E., (2019). Dye residues in aquaculture products: Targeted and metabolomics mass spectrometric approaches to track their abuse. *Food Chem., 294*, 355–367.

Espinoza, E. O., Lancaster, C. A., Kreitals, N. M., Hata, M., Cody, R. B., & Blanchette, R. A., (2014). Distinguishing wild from cultivated agarwood (*Aquilaria* spp.) using direct analysis in real-time and time-of-flight mass spectrometry. *Rapid Commun Mass Spectrom., 28*, 281–289.

Fibigr, J., Šatínský, D., & Solich, P., (2018). Current trends in the analysis and quality control of food supplements based on plant extracts. *Anal. Chim. Acta, 1036*, 1–15.

Finch, K., Espinoza, E., Jones, F. A., & Cronn, R., (2017). Source identification of western Oregon Douglas-fir wood cores using mass spectrometry and random forest classification. *Appl. Plant Sci., 5*, 1600158.

Fu, Z., Li, Z., Hu, P., Feng, Q., Xue, R., Hu, Y., & Huang, C., (2015). A practical method for the rapid detection and structural characterization of major constituents from traditional Chinese medical formulas: Analysis of multiple constituents in yinchenhao decoction. *Anal. Methods., 7*, 4678–4690.

Giraudo, A., Grassi, S., Savorani, F., Gavoci, G., Casiraghi, E., & Geobaldo, F., (2019). Determination of the geographical origin of green coffee beans using NIR spectroscopy and multivariate data analysis. *Food Control., 99*, 137–145.

Goessens, T., Huysman, S., De Troyer, N., Deknock, A., Goethals, P., Lens, L., Vanhaecke, L., & Croubels, S., (2020). Multi-class analysis of 46 antimicrobial drug residues in pond water using UHPLC-Orbitrap-HRMS and application to freshwater ponds in Flanders, Belgium. *Talanta, 220*, 121326.

Gómez-Ramos, M. M., Ferrer, C., Malato, O., Agüera, A., & Fernández-Alba, A. R., (2013). Liquid chromatography-high-resolution mass spectrometry for pesticide residue analysis in fruit and vegetables: Screening and quantitative studies. *J. Chromatogr. A., 1287*, 24–37.

Grund, B., Marvin, L., & Rochat, B., (2016). Quantitative performance of a quadrupole-orbitrap-MS in targeted LC-MS determinations of small molecules. *J. Pharm. Biomed. Anal., 124*, 48–56.

Guerreiro, T. M., De Oliveira, D. N., Melo, C. F. O. R., De Oliveira, L. E., Da Silva, R. M., & Catharino, R. R., (2018). Evaluating the effects of the adulterants in milk using direct-infusion high-resolution mass spectrometry. *Food Res. Int., 108*, 498–504.

Hahner, S., (2002). Analysis of short tandem repeat polymorphisms by electrospray ion trap mass spectrometry. *Nucleic Acids Res., 28*, 82e.

Hofstadler, S. A., Sannes-Lowery, K. A., & Hannis, J. C., (2005). Analysis of nucleic acids by FTICR MS. *Mass Spectrom. Rev., 24*, 265–285.

Hollender, J., Schymanski, E. L., Singer, H. P., & Ferguson, P. L., (2017). Nontarget screening with high resolution mass spectrometry in the environment: Ready to go? *Environ. Sci. Technol., 51*, 11505–11512.

Hrbek, V., Vaclavik, L., Elich, O., & Hajslova, J., (2014). Authentication of milk and milk-based foods by direct analysis in real-time ionization–high resolution mass spectrometry (DART–HRMS) technique: A critical assessment. *Food Control, 36*, 138–145.

HUANG, X. M., Yang, W., Jun-Tao, C., Fu-Hua, W., Xu, W., Ya-Fei, L., & Wei-Yun, W., (2019). Applications of high-resolution mass spectrometry in the determination of chlorinated paraffins. *Chin J Anal Chem., 47*, 323–334.

Innamorato, V., Longobardi, F., Cervellieri, S., Cefola, M., Pace, B., Capotorto, I., Gallo, V., et al., (2020). Quality evaluation of table grapes during storage by using 1H NMR, LC-HRMS, MS-eNose and multivariate statistical analysis. *Food Chem., 315*, 126247.

ISO 9000:2015, (2015). *Quality Management Systems: Fundamentals and Vocabulary*. ISO, Geneva.

Jahnke, A., Gandrass, J., & Ruck, W., (2004). Simultaneous determination of alkylphenol ethoxylates and their biotransformation products by liquid chromatography/electrospray ionization tandem mass spectrometry. *J. Chromatogr. A., 1035*, 115–122.

JEOL USA Inc., (2006). *Flavones and Flavor Components in Two Basil Leaf Chemotypes*. JEOL application note.

Jessika, H., (2009). Analysis of brominated dioxins and furans by high resolution gas chromatography/high resolution mass spectrometry. *J. Chromatogr. A, 1216*, 376–384.

Kaklamanos, G., Vincent, U., & Von, H. C., (2013). Multi-residue method for the detection of veterinary drugs in distillers grains by liquid chromatography–Orbitrap high resolution mass spectrometry. *J. Chromatogr. A., 1322*, 38–48.

Kim, H. J., Baek, W. S., & Jang, Y. P., (2011). Identification of ambiguous cubeb fruit by DART-MS-based fingerprinting combined with principal component analysis. *Food Chem., 129*, 1305–1310.

Kim, H. J., Seo, Y. T., Park, S. I., Jeong, S. H., Kim, M. K., & Jang, Y. P., (2015). DART-TOF-MS based metabolomics study for the discrimination analysis of geographical origin of Angelica gigas roots collected from Korea and China. *Metabolomics, 11*, 64–70.

Krahmer, M. T., Johnson, Y. A., Walters, J. J., Fox, K. F., Fox, A., & Nagpal, M., (1999). Electrospray quadrupole mass spectrometry analysis of model oligonucleotides and polymerase chain reaction products: Determination of base substitutions, nucleotide additions/deletions, and chemical modifications. *Anal. Chem., 71*, 2893–2900.

Lesiak, A. D., & Musah, R. A., (2016). More than just heat: Ambient ionization mass spectrometry for determination of the species of origin of processed commercial products–application to psychoactive pepper supplements. *Anal. Methods, 8*, 1646–1658.

Li, Y., Zhang, J., Jin, Y., Wang, L., Zhao, W., Zhang, W., Zhai, L., et al., (2016). Hybrid quadrupole-orbitrap mass spectrometry analysis with accurate-mass database and parallel reaction monitoring for high throughput screening and quantification of multi-xenobiotics in honey. *J. Chromatogr. A, 1429*, 119–126.

Lippolis, V., De Angelis, E., Fiorino, G. M., Di Gioia, A., Arlorio, M., Logrieco, A. F., & Monaci, L., (2020). Geographical origin discrimination of monofloral honeys by direct analysis in real Time ionization-high resolution mass spectrometry (DART-HRMS). *Foods, 9*, 1205.

López-Gutiérrez, N., Romero-González, R., Vidal, J. L. M., & Frenich, A. G., (2016). Quality control evaluation of nutraceutical products from Ginkgo biloba using liquid chromatography coupled to high resolution mass spectrometry. *J. Pharm. Biomed Anal., 121*, 151–160.

López-Ruiz, R., Romero-González, R., & Frenich, A. G., (2019). Metabolomics approaches for the determination of multiple contaminants in food. *Curr. Opin. Food Sci., 28*, 49–57.

Luo, X., Zhang, L., Niu, Z., Ye, X., Tang, Z., & Xia, S., (2017). Liquid chromatography coupled to quadrupole-Orbitrap high resolution mass spectrometry based method for target analysis and suspect screening of non-ionic surfactants in textiles. *J. Chromatogr. A, 1530*, 80–89.

Maleknia, S. D., Vail, T. M., Cody, R. B., Sparkman, D. O., Bell, T. L., & Adams, M. A., (2009). Temperature-dependent release of volatile organic compounds of eucalypts by direct analysis in real-time (DART) mass spectrometry. *Rapid Commun. Mass Spectrom., 23*, 2241–2246.

Marchal, A., Prida, A., & Dubourdieu, D., (2016). A new approach for differentiating sessile and pedunculate oak: Development of a LC-HRMS method to quantify triterpenoids in wood. *J. Agric. Food Chem., 64*, 618–626.

Martínez-Domínguez, G., Romero-González, R., & Frenich, A. G., (2016). Multi-class methodology to determine pesticides and mycotoxins in green tea and royal jelly supplements by liquid chromatography coupled to Orbitrap high resolution mass spectrometry. *Food Chem., 197*, 907–915.

Martínez-Domínguez, G., Romero-González, R., Arrebola, F. J., & Frenich, A. G., (2016). Multi-class determination of pesticides and mycotoxins in isoflavones supplements obtained from soy by liquid chromatography coupled to Orbitrap high resolution mass spectrometry. *Food Control, 59*, 218–224.

Masi, E., Taiti, C., Heimler, D., Vignolini, P., Romani, A., & Mancuso, S., (2016). PTR-TOF-MS and HPLC analysis in the characterization of saffron (*Crocus sativus* L.) from Italy and Iran. *Food Chem., 192*, 75–81.

Minaei, S., Shafiee, S., Polder, G., Moghadam-Charkari, N., Van, R. S., Barzegar, M., Zahiri, J., et al., (2017). VIS/NIR imaging application for honey floral origin determination. *Infrared Phys. Technol., 86*, 218–225.

Müller, G., Norpoth, K., & Travenius, S. Z. M., (1981). Quantitative determination of bis (chloromethyl) ether (BCME) in the ppb range by using portable air sample collectors. *Int. Arch Occup. Environ. Health, 48*, 325–329.

Musah, R. A., Espinoza, E. O., Cody, R. B., Lesiak, A. D., Christensen, E. D., Moore, H. E., Maleknia, S., & Drijfhout, F. P., (2015). A high throughput ambient mass spectrometric approach to species identification and classification from chemical fingerprint signatures. *Sci. Rep., 5*, 11520.

Navarro, M., Núñez, O., Saurina, J., Hernández-Cassou, S., & Puignou, L., (2014). Characterization of fruit products by capillary zone electrophoresis and liquid chromatography using the compositional profiles of polyphenols: Application to authentication of natural extracts. *J. Agric. Food Chem., 62*, 1038–1046.

Neu, V., Bielow, C., Gostomski, I., Wintringer, R., Braun, R., Reinert, K., Schneider, P., Stuppner, H., & Huber, C. G., (2013). Rapid and comprehensive impurity profiling of synthetic thyroxine by ultrahigh-performance liquid chromatography–high-resolution mass spectrometry. *Anal. Chem., 85*, 3309–3317.

Neumann, E. K., Comi, T. J., Spegazzini, N., Mitchell, J. W., Rubakhin, S. S., Gillette, M. U., Bhargava, R., & Sweedler, J. V., (2018). Multimodal chemical analysis of the brain by high mass resolution mass spectrometry and infrared spectroscopic imaging. *Anal. Chem., 90*, 11572–11580.

Nunes, V. O., Silva, R. V., Romeiro, G. A., & Azevedo, D. A., (2020). The speciation of the organic compounds of slow pyrolysis bio-oils from Brazilian tropical seed cake fruits using high-resolution techniques: GC× GC-TOFMS and ESI (±)-orbitrap HRMS. *Microchem. J., 153*, 104514.

Orr, A., Stotesbury, T., Wilson, P., & Stock, N. L., (2019). The use of high-resolution mass spectrometry (HRMS) for the analysis of DNA and other macromolecules: A how-to guide for forensic chemistry. *Forensic Chem., 14*, 100169.

Pandey, R., Chandra, P., Kumar, B., Srivastva, M., Aravind, A. A., Shameer, P. S., & Rameshkumar, K. B., (2015). Simultaneous determination of multi-class bioactive constituents for quality assessment of *Garcinia* species using UHPLC–QqQLIT–MS/MS. *Ind. Crops Prod., 77*, 861–872.

Peng, J., Wu, D., Jiang, Y., Zhang, J., Lin, X., Lu, S., Han, P., et al., (2018). Spatiotemporal variability of polybrominated diphenyl ether concentration in atmospheric fine particles in Shenzhen, China. *Environ. Pollut., 238*, 749–759.

Périat, A., Bieri, S., & Mottier, N., (2019). SWATH-MS screening strategy for the determination of food dyes in spices by UHPLC-HRMS. *Food Chem. X, 1*, 100009.

Pico, Y., Alfarhan, A. H., & Barcelo, D., (2020). How recent innovations in gas chromatography-mass spectrometry have improved pesticide residue determination: An alternative technique to be in your radar. *TrAC Trends Anal. Chem., 122*, 115720.

Pizzini, S., Marchiori, E., Piazza, R., Cozzi, G., & Barbante, C., (2015). Determination by HRGC/HRMS of PBDE levels in edible Mediterranean bivalves collected from north-western Adriatic coasts. *Microchem. J., 121*, 184–191.

Pontes, N. S., Silva, R. V., Ximenes, V. L., Pinho, A. R., & Azevedo, D. A., (2020). Chemical speciation of petroleum and bio-oil coprocessing products: Investigating the introduction of renewable molecules in refining processes. *Fuel, 288*, 119654.

Qiao, X., Song, W., Wang, Q., Zhang, Z. X., Bo, T., Li, R. Y., Liang, L. N., et al., (2015). Comprehensive chemical analysis of triterpenoids and polysaccharides in the medicinal mushroom *Antrodia cinnamon*. *RSC Adv., 5*, 47040–47052.

Raclariu, A. C., Paltinean, R., Vlase, L., Labarre, A., Manzanilla, V., Ichim, M. C., Crisan, G., et al., (2017). Comparative authentication of *Hypericum perforatum* herbal products using DNA metabarcoding, TLC and HPLC-MS. *Sci Rep., 7*, 1–12.

Raclariu, A. C., Ţebrencu, C. E., Ichim, M. C., Ciupercă, O. T., Brysting, A. K., & De Boer, H., (2018). What's in the box? Authentication of Echinacea herbal products using DNA metabarcoding and HPTLC. *Phytomedicine, 44*, 32–38.

Ribeiro, M. A., Murgu, M., De Moraes, S. V., Sawaya, A. C., Ribeiro, L. F., Justi, A., & Meurer, E. C., (2017). The screening of organic matter in mineral and tap water by UHPLC-HRMS. *Talanta, 174*, 581–586.

Riuzzi, G., Tata, A., Massaro, A., Bisutti, V., Lanza, I., Contiero, B., Bragolusi, M., et al., (2021). Authentication of forage-based milk by mid-level data fusion of (+/−) DART-HRMS signatures. *Int. Dairy J., 112*, 104859.

Rivera-Pérez, A., López-Ruiz, R., Romero-González, R., & Frenich, A. G., (2020). A new strategy based on gas chromatography–high resolution mass spectrometry

(GC–HRMS-Q-Orbitrap) for the determination of alkenyl benzenes in pepper and its varieties. *Food Chem., 321*, 126727.

Rovira, J., & Domingo, J. L., (2019). Human health risks due to exposure to inorganic and organic chemicals from textiles: A review. *Environ. Res., 168*, 62–69.

Salunkhe, D. K., & Desai, B. B., (1988). Effects of agricultural practices, handling, processing, and storage on vegetables. In: Karmas, E., & Harris, R. S., (eds.), *Nutritional Evaluation of Food Processing* (pp. 23–71). Springer: Dordrecht.

Schnaible, V., Wefing, S., Resemann, A., Suckau, D., Bücker, A., Wolf-Kümmeth, S., & Hoffmann, D., (2002). Screening for disulfide bonds in proteins by MALDI in-source decay and LIFT-TOF/TOF-MS. *Anal. Chem., 74*, 4980–4988.

Schuster, C., Paal, M., Lindner, J., Zoller, M., Liebchen, U., Scharf, C., & Vogeser, M., (2019). Isotope dilution LC-orbitrap-HRMS with automated sample preparation for the simultaneous quantification of 11 antimycotics in human serum. *J. Pharm. Biomed. Anal., 166*, 398–405.

Schwarzenberg, A., Dossmann, H., Cole, R. B., Machuron-Mandard, X., & Tabet, J. C., (2014). Differentiation of isomeric dinitrotoluenes and aminodinitrotoluenes using electrospray high resolution mass spectrometry. *J. Mass Spectrom., 49*, 1330–1337.

Shi, F., Guo, C., Gong, L., Li, J., Dong, P., Zhang, J., Cui, P., et al., (2014). Application of a high resolution benchtop quadrupole-Orbitrap mass spectrometry for the rapid screening, confirmation and quantification of illegal adulterated phosphodiesterase-5 inhibitors in herbal medicines and dietary supplements. *J. Chromatogr. A., 1344*, 91–98.

Shin, J. H., & Baek, Y. J., (2012). Analysis of polybrominated diphenyl ethers in textiles treated by brominated flame retardants. *Text. Res. J., 82,* 1307–1316.

Sokolowska, I., Mo, J., Rahimi, P. F., McVean, C., Meijer, L. A., Switzar, L., Balog, C., et al., (2019). Implementation of a high-resolution liquid chromatography-mass spectrometry method in quality control laboratories for release and stability testing of a commercial antibody product. *Anal. Chem., 92*, 2369–2373.

Solouki, T., Emmett, M. R., Guan, S., & Marshall, A. G., (1997). Detection, number, and sequence location of sulfur-containing amino acids and disulfide bridges in peptides by ultrahigh-resolution MALDI FTICR mass spectrometry. *Anal. Chem., 69*, 1163–1168.

Souard, F., Delporte, C., Stoffelen, P., Thévenot, E. A., Noret, N., Dauvergne, B., Kauffmann, J. M., et al., (2018). Metabolomics fingerprint of coffee species determined by untargeted-profiling study using LC-HRMS. *Food Chem., 245*, 603–612.

Stella, R., Sette, G., Moressa, A., Gallina, A., Aloisi, A. M., Angeletti, R., & Biancotto, G., (2020). LC-HRMS/MS for the simultaneous determination of four allergens in fish and swine food products. *Food Chemistry, 331*, 127276.

Su, C., Li, C., Sun, K., Li, W., & Liu, R., (2020). Quantitative analysis of bioactive components in walnut leaves by UHPLC-Q-orbitrap HRMS combined with QAMS. *Food Chem., 331*, 127180.

Sugimoto, M., Obiya, S., Kaneko, M., Enomoto, A., Honma, M., Wakayama, M., Soga, T., & Tomita, M., (2017). Metabolomic profiling as a possible reverse engineering tool for estimating processing conditions of dry-cured hams. *J. Agric. Food Chem., 65*, 402–410.

Temerdashev, A., Azaryan, A., & Dmitrieva, E., (2020). Meldonium determination in milk and meat through UHPLC-HRMS. *Heliyon, 6*, e04771.

Van, D. B. M., Denison, M. S., Birnbaum, L. S., Devito, M. J., Fiedler, H., Falandysz, J., Rose, M., et al., (2013). Polybrominated dibenzo-p-dioxins, dibenzofurans, and biphenyls: Inclusion in the toxicity equivalency factor concept for dioxin-like compounds. *Toxicol. Sci., 133*, 197–208.

Van, L. J. A., Vervoort, J., Stroomberg, G., & De Voogt, P., (2014). Identification of unknown microcontaminants in Dutch river water by liquid chromatography-high resolution mass spectrometry and nuclear magnetic resonance spectroscopy. *Environ. Sci. Technol., 48*, 12791–12799.

Vichi, S., Jerí, Y., Cortés-Francisco, N., Palacios, O., & Caixach, J., (2014). Determination of volatile thiols in roasted coffee by derivatization and liquid chromatography–high resolution mass spectrometric analysis. *Food Res. Int., 64*, 610–617.

Wang, X. B., Zhenga, J., Li, J. J., Yu, H. Y., Li, Q. Y., Xu, L. H., Liu, M. J., et al., (2018). Simultaneous analysis of 23 illegal adulterated aphrodisiac chemical ingredients in health foods and Chinese traditional patent medicines by ultrahigh performance liquid chromatography coupled with quadrupole time-of-flight mass spectrometry. *J. Food Drug Anal., 26*, 1138–1153.

Whitehead, A. J., (1998). Ensuring food quality and safety and FAO technical assistance. *Food, Nutrition and Agriculture, 21*, 10–17.

Yao, C., Yang, W., Si, W., Pan, H., Qiu, S., Wu, J., Shi, X., et al., (2016). A strategy for establishment of practical 1 identification methods for Chinese patent medicine from systematic multi-component characterization to selective ion monitoring of chemical markers: Shuxiong tablet as a case study. *RSC Adv., 6*, 65055–65066.

Yu, H., An, Y., Battistel, M. D., Cipollo, J. F., & Freedberg, D. I., (2018). Improving analytical characterization of glycoconjugate vaccines through combined high-resolution MS and NMR: Application to *Neisseria* meningitidis serogroup B oligosaccharide-peptide glycoconjugates. *Anal. Chem., 90*, 5040–5047.

Zeng, K., Geerlof-Vidavisky, I., Gucinski, A., Jiang, X., & Boyne, M. T., (2015). Liquid chromatography-high resolution mass spectrometry for peptide drug quality control. *AAPS J., 17*, 643–651.

Zeng, X., Zheng, Y., Luo, J., Liu, H., & Su, W., (2020). A review on the chemical profiles, quality control, pharmacokinetic and pharmacological properties of fufang xueshuantong capsule. *J. Ethnopharmacol.*, 113472.

Zhan, F., Zhang, H., Cao, R., Fan, Y., Xu, P., & Chen, J., (2019). Release and transformation of BTBPE during the thermal treatment of flame retardant ABS plastics. *Environ. Sci. Technol., 53*, 185–193.

Zhang, L., Tian, K., Wang, Y., Zou, J., & Du, Z., (2017). Characterization of ancient Chinese textiles by ultra-high-performance liquid chromatography/quadrupole-time of flight mass spectrometry. *Int. J. Mass Spectrom., 421*, 61–70.

Zhang, Y., Zhao, Y., Shen, G., Zhong, S., & Feng, J., (2018). NMR spectroscopy in conjugation with multivariate statistical analysis for distinguishing plant origin of edible oils. *J. Food Compost. Anal., 69*, 140–148.

Zhu, L., Yang, S., Li, G., Zhang, X., Yang, J., Lai, X., & Yang, G., (2016). Simultaneous analysis of tocopherols, tocotrienols, phospholipids, γ-oryzanols and β-carotene in rice by ultra-high performance liquid chromatography coupled to linear ion trap-orbitrap mass spectrometer. *Anal. Methods., 8*, 5628–5637.

Zomer, P., & Mol, H. G. J., (2015). Simultaneous quantitative determination, identification and qualitative screening of pesticides in fruits and vegetables using LC-Qorbitrap-MS. *Food Addit. Contam. Part A, 32*, 1628–1636.

Zuo, T., Qian, Y., Zhang, C., Wei, Y., Wang, X., Wang, H., Hu, Y., et al., (2019). Data-dependent acquisition and database-driven efficient peak annotation for the comprehensive profiling and characterization of the multicomponent from compound xueshuantong capsule by UHPLC/IM-QTOF-MS. *Molecules, 24*, 3431.

CHAPTER 6

HRMS for Forensic Studies and Investigations

AKHILA NAIR,[1,2] JÓZEF T. HAPONIUK,[1] and SREERAJ GOPI[1]

[1]*Department of Chemistry, Gdansk University of Technology, Gdansk, Poland, E-mail: sreerajgopi@yahoo.com (S. Gopi)*

[2]*R&D Center, Aurea Biolabs (P) Ltd., Kolenchery, Cochin – 682311, Ernakulam, Kerala, India*

ABSTRACT

Forensic science forms a prime sector of law-enforcing bodies to supply proofs related to possible crime. The forensic scientist searches numerous materials collected by police from the place of the crime that conclude to the cause of the crime or to the suspect. The test matters are sometimes illegal drugs from dead bodies, or body fluids, residues from explosives, poison in cadavers or dyes of ink in forged documents. High resolution mass spectrometry (HRMS) forms the main tool in forensic science to get unbiased results with high resolution (HR), accuracy, precision, selectivity and sensitivity. This chapter deals with the comprehensive application of the LCMS or HRMS in forensic investigation for the comprehension of the mode by which forensic scientists analyze the samples with these instruments that help resolve any crime case.

High-Resolution Mass Spectrometry and Its Diverse Applications: Cutting-Edge Techniques and Instrumentation. Sreeraj Gopi, PhD, Sabu Thomas, PhD, Augustine Amalraj, PhD, & Shintu Jude, MSc (Editors)

6.1 INTRODUCTION

The term forensic delineate used in, belonging to or suitable for court and forensic science is the science which describes the queries of law. This science collaborates with geology, biology, chemistry, and computer science to understand, detect, and identify the evidence from a crime scene. The fingerprints, DNA analysis, detection of trace elements from a crime scene are the major approach to solve the questions related to a crime. The results of these analyzes become the prime sources that help channel any investigation and clarify any perplexing problem (Morgan, 2019). The reliability of evidence is the major jurisdiction challenges encountered by the court which can be subdued by the collecting and observing false positive error rates to guarantee a genuine value (Martire ct al., 2019). In addition, employing advanced and sophisticated techniques, namely high-resolution mass spectrometry (HRMS) are helpful measures which could solve critical cases.

HRMS is an upcoming technique with a broadband resolving power greater than 10,000. There are various types of HRMS such as time of flight (TOF), Fourier transform-ion cyclotron resonance-mass spectrometer (FT-ICR-MS), sector mass spectrometer (SECTOR) and Fourier transform orbitrap (FT Orbitrap), which differ by the mass analyzer or detector structure used. Most commonly used mass analyzers are Orbitrap and TOF. However, these mass analyzers, if used solely possess certain problems like weak multi-stage MS and low sensitivity. That tends to circumscribe their applications. Therefore, the concept of tandem high-resolution mass spectrometry (THRMS) materialized which utilized ion-trap (IT)-TOF, quadrupole (Q)-TQF), Q Exactive-Orbitrap and LTQ-Orbitrap to provide high resolution (HR) mass spectra and accurate monoisotopic mass measurements. Besides, HRMS is also utilized by coupling them with gas chromatography (GC), ultra-high-performance liquid chromatography (UHPLC) or high-performance liquid chromatography (HPLC) to achieve results with utmost precision. The resolution received is dependent upon molecular weight and elemental composition of the compound subjected to analysis (Lin et al., 2015). It is observed that a resolution of 0.1 mDa is reasonably quite enough for mass assessment for compounds having a mass range from 0 to 500 Da (Lin, 2015).

HRMS has far-reaching application in the field of food safety, medicine, bioengineering, chemical engineering, environmental protection and forensic because it provides a broad mass range, high sensitivity, high quality resolution, full scanning speed. In forensic science it has been identified as a strong candidate for its efficiency in analyzing blood, urine, hair, soil, ink, poison, and explosives samples.

6.2 DRUG IDENTIFICATION IN BLOOD, URINE, HAIR SAMPLES

In certain cases, samples from the larvae, reared on putrefied bodies which are in least quantity are successfully detected by liquid chromatography-tandem mass spectrometry (LC-MS). Δ^9-tetrahydrocannabinol (THC) and its metabolite 11-carboxy-delta-9-tetrahydrocannabinol (THCA): compounds bioaccumulated in carrion feeding insects are simultaneously detected effortlessly by LC-MS in an investigation by Karampela et al. In this study, multiple extraction techniques were employed for the extraction of these compounds which are detected by negative mode electrospray ionization source with selected ions recognized at m/z 313 and m/z 343 for THC and THCA respectively. Thus, these findings were a great breakthrough in many toxicological cases (Karampela et al., 2015). Another study demonstrated that a suitable method for distinguishing between isomers of Δ^9-THC and cannabidiol (CBD) with identical fragment spectra, which is difficult to identify. Therefore, Hydrogen/deuterium (H/D) exchange experiments by utilizing hybrid flow injection-quadrupole-time-of-flight mass spectrometry (FI-TOF-MS) with acid-catalyzed in-source equilibrium among CBD and THC was set to form same precursor ion and accelerated acidic condition in positive ESI droplets lead to H/D exchange in methyl group. For this, CBD and THC solutions were prepared in D_2O/acetonitrile (50:50, v/v) mobile phase compared the same measurements of the compounds in H_2O with 0.5% formic acid (negative ESI mode) which were analyzed by FI-QTOF-MS. For both preparations, the position and the number of hydrogen atoms exchanged were elucidated at 10, 20, and 40 eV for the MS and collision induced dissociation (CID) spectra. Around 7 to 8 hydrogen atoms were interchanged by deuterium in THC and CBD. Precursors available in positive ESI CID spectra exchanged 4 D-atoms and carbon atoms of the molecules situated in the non-aromatic part underwent H/D exchange apart from the OH group for CBD and THC. However, in negative ESI mode, there were no H/D exchange with the addition of OH group was found and different CID spectra for both THC and CBD were interpreted. It is also suggested that positive LC/ESI-MS/MS for identification of peak by CID spectra or an abundance ratio of multiple reaction monitoring (MRM) falls insufficient to distinguishing between CBD and THC and investigated substances in D_2O are promising to perform H/D exchange experiments along with the incorporation of retention time (RT) for measurement by FI-QTOF-MS (Broecker, 2012).

Besides, determination of atropine and scopolamine in human plasma (100 L) was optimized by using LC-MS/MS. In this procedure the sample pre-treatment was carried out by protein precipitation with acetonitrile and thereafter concentration. Also, levobupivacaine, used as internal standard (IS) as well as the analytes were separated by using column Zorbax XDB-CN having parameters 75 mm x 46 mm x 3.5 mm, mobile phase-purified water, acetonitrile, and formic acid fabricated in gradient elution system. The method utilized matrix effect in ESI positive mode for analysis in triple quadrupole MS for deproteinized plasma samples as well as selected reaction monitoring (SRM) was utilized for quantifying in the range of 0.10–50.00 ng/mL. At lower limit of quantification (LLOQ) the interday precision for tropanes and atropine were less than 10%, scopolamine less than 14% and less than 18%. It was observed that the mean intraday and interday accuracies for scopolamine and atropine was in the range of +11% and +7%, respectively. Hence, the method proved very efficient for investigating the toxic and therapeutic level of both compounds (Koželj, 2014). Besides, constitutional isomers have similar structure due to which their identification poses a tremendous problem as each isomer possess different pharmacological and toxicological effects and therefore their identification is very essential. In this context, a method was developed by using LC-MS/MS having a core-shell bi-phenyl analytical column for the identification of para, meta, and ortho isomers of methylethcathione (MEC) and methyl methcathinone (MMC). The method developed for their estimation was validated for their reliability with linearity in the range between 5 ng/mL and 250 ng/mL and LLOQ as well as LOD was found to be 5 and 2 ng/mL respectively. With this method, 8 cases were studied, among them MMC isomers were found in three cases, where one was observed positive for 2-MMC and other two were found positive for 3-MMC and none were detected with 4-MMC, 2-MEC and 4-MEC. These observations facilitated quick response to new recreational drugs and the wavering drug laws (Maas et al., 2016). Zolpidem, a nonbenzodiazepine drug is termed as a date-rape drug and narcotics substitute around the globe because it induces sleep. Ingestion of this drug reported as one of the causes of road accidents and has been placed under drug-facilitated sexual assault (DFSA) by society of forensic toxicologists. The quantitative and qualitative method for analyzing this drug and their metabolites in blood and oral fluid (OF) was effectively done through LC/MS by adopting polymeric ion-exchange sorbent in solid-phase extraction (SPE) for preparation method for samples. Obtained method exhibited good linearity with determination coefficient (R^2) of 0.9998 and 0.9989 for OF and blood samples respectively. LOQ and LOD of zolpidem were 0.2 ng/mL for blood and 1.0 ng/mL for OF samples. SPE method recovery varied from 79.9 ± 12.6 mL to 104.1± 1.77 mL for blood sample and 80.2 ± 0.48 mL to 103.8 ± 1.51 mL for OF sample. The samples collected from zolpidem taken patients were analyzed.

After 15 hrs., zolpidem was detected in patients who ingested 10 mg of zolpidem tartrate from both the variety of samples. This method allowed quantification of zolpidem in subtherapeutic range, therapeutics range, and qualitatively of their main metabolites in both OF and blood samples. This method qualifies the criteria necessary for forensic and clinical investigation (Piotrowski et al., 2015). For the reduction of the backlog and to contribute to more timely prosecution of criminals, the forensic department opted for Helium DART-TOFMS (He-DART-TOFMS) over a decade ago. Current developments in these high resolving power instruments took the shape of N_2 DART-TOFMS (using nitrogen) that could efficiently ionize all polar compounds except for acetonitrile and methanol and would be helpful in rapid screening of forensic drugs. This N_2 DART-TOFMS utilized ionsense DART-100 as ion source and JEOL AccuTOF mass spectrometer to analyze samples within 3 minutes. The 1 μL samples were introduced by a closed-end capillary tube into the nitrogen stream which was between the orifice of AccuTOF and DART-100. LOD was about 10 μg/mL or 10 pg and mass errors were less than 5 mDa for the forensic drugs (10 abused drugs, 4 prescription drugs and 8 synthetic drugs) which were positively identified by $[M+H]^+$ ions, characteristic N_2 DART ions and in-source fragmentation which are not able to be generated by Helium DART like $[M+H+2O]^+$, $[M+H+O]^+$. Besides, N_2 is more reliable, cheap, convenient, and readily available in air than H_2. Hence, N_2 DART-TOFMS is a better alternative for analyzing drugs under forensic investigation (Song et al., 2019).

Most immune screening methods employed for urine and serum samples in forensic investigation are very expensive, and detected only certain drugs like opiates, cannabinoids, methadone, cocaine (COC), amphetamines (AP), and benzodiazepines with the influence of matrix effects and cross-reactivity, which affected the overall results and hence, were not considered a proof during court cases. Thus, a requirement of the efficient screening method was required, which led Dziadosz and coworkers to develop LC-MS/MS based screening method. The utilization of two mass spectrometers with different sensitivities were used for method validation. Sample volume 100 μL with C-18 Luna column, 5 μm, 100 A, 150×2 mm and mobile phase consisting of two different ratios of water and methanol, one consisting of 95:5 and other 3:97, both having 0.1% acetic acid and 10 mM ammonium acetate were successfully utilized for qualitative analyzes which were relatively less costly than earlier methods. They were able to use simple precipitation method for direct analysis of a broad spectrum of toxicological drugs which is indeed a relief from a forensic point of view (Dziadosz et al., 2018). Another drug of abuse is synthetic cannabinoids, which have similar activity as endocannabinoid. However, their mixtures having chemical coatings which are sold under name legal or natural products over the internet. The contents of these products, especially JWH-073, JWH-073 N-(4-Hydroxybutyl), JWH-073 N-butanoic acid are successfully quantified and detected by UPLC-MS/MS in urine and blood samples.

The method validation report suggested linearity was about 0.1–50 ng/mL, precision, selectivity, intra assay accuracy, intra-assay was found to be less than 10%, recovery within 75–95%, LOQs in the range of 0.11–0.17 ng/mL, limit of detection 8 ng/mL to 0.13 ng/mL. In addition, stability, matrix effects and process efficiency were also monitored, providing satisfactory results (Ozturk et al., 2015). Apart from JWH-073, JWH-018, AM-2201, XLR-11, UR-144, MAM-2201, JWH-210, AKB-48, JWH-122, PB-22 and a-PVP were detected from the same samples via LC-MS/MS and GC-MS. The detection by LC-MS/MS was carried out by using an AB Sciex QTRAP1 5500 triple, quadruple linear ion trap mass spectrometer instrument and a Restek ultra biphenyl having column dimensions –150 mm × 2.1 mm × 3 mm for chromatographic separations. The MS/MS detection was accomplished utilizing negative and positive electrospray ionization in MRM mode and injection volumes were 10 and 20 μL, respectively (Ozturk et al., 2015). Similarly, Paul et al. studied the various stimulant designer drugs such as AP, cathinone, phenethylamine, and piperazine derivatives as well as drug of abuse like ritalinic acid and ketamine (KET) in urine samples with the help of LC-HRMS-QTOF. Besides this, the sample preparation was supported by a rapid and convenient procedure named salting-out liquid-liquid extraction (SALLE) followed by method validation including selectivity, accuracy, linearity, stability, LOQ, LOQ, recoveries, matrix effects for 39 compounds out of which 35 were under acceptable quantitative range. Moreover, an untargeted data dependent acquisition mode favored metabolite identification of methylenedioxypyrovalerone (MDPV) and KET without reference standards. These observations showcased that this strategy is suitable for untargeted analysis and combined targeted drug of abuse in the given biological matrix-like urine (Paul et al., 2014). The determination of biological matrices of lysergic acid diethylamide (LSD) poses a great problem for the forensic toxicologist because they are administered in very low quantity. A reliable, sensitive method developed by using LC-MS/MS proved a relief for the forensic department in the analysis of LSD and 2-oxo-3-hydroxy-LSD (O-H-LSD), its metabolite in urine and hair samples. A separate method of extraction was utilized for both urine and hair samples. Urine samples were subjected to LLE and hair samples were retrieved by using methanol for 15 hrs. at 38°C. It was observed that the LOD was 0.25 pg/mg for LSD and 0.5 pg/mL for O-H-LSD in hair and 0.01 ng/mL for LSD and 0.025 ng/mL for O-H-LSD with desirable linearity, precision, and accuracy.

Two cases were examined for LSD and O-H-LSD where LSD were found to be 1.27 pg/mL and 0.95 pg/mg and O-H-LSD was observed only in one case with values lower than LOQ in hair samples. However, for urine samples LSD were about 0.48 ng/mL and 2.70 ng/mL and O-H-LSD were found to be 4.19 ng/mL and 25.2 ng/mL in each case (Jang et al., 2015). Besides, Favretto et al. (2012) also reported that the concurrent use of HRMS full scan and source collisional experiments laid the foundation of retrospective analysis of untargeted drugs, especially drug of abuse and pharmaceuticals in 2.5 mg hair samples. They have screened, quantified, and validated 28 drugs and their metabolites such as COC, opioids, AP, hallucinogens, benzodiazepines, and antidepressants. The hair samples were decontaminated and extracted by 200 μL of mixture in the ratio 80:10:10, v/v (water, acetonitrile, and 1 M trifluoroacetic acid) with 5 min pulverization or extraction step. They analyzed these samples by HPLC-HRMS in an Orbitrap with a nominal resolution of 60,000 and run time 26 min. The linearity ranged from 0.9964 to 0.9999, LOQ was 0.1 ng/mg (8 compounds), 0.5 ng/mg (4 compounds) and 0.2 ng/mg (16 compounds). Intermediate reproducibility as well as repeatability was less than 20% with better recoveries (75%) and process efficiency (65%) (Favretto et al., 2012). Miyaguchi et al. demonstrated that utilization of LTQ-Orbitrap-XL hybrid mass spectrometry is useful for forensic department for the analysis of illicit drugs such as AP, dimethylaamphetamine, methamphetamine (MA), 3,4-methylenedioxyamphetamine (MDA), 3,4-methylenedioxymethamphetamine (MDMA), norketamine (NK), KET, COC, benzoylecgonine (BE) in hair samples. Their study utilized micro pulverized method for sample preparation and found good recoveries of COC (35.5%) and AP (71.7%). LLOQ was found to be 0.50 ng mg^{-1} (MDA), 0.10 ng mg^{-1} (AP, COC, MA, DMA, NK) and 0.050 ng mg^{-1} (BE, KET, MDMA), for 0.2 mg of a hair sample. For quantification, a tolerance of 3 ppm was set for HR extracted ion chromatograms and both the analyzers utilized, namely linear IT analyzer and Orbitrap analyzer obtained unit resolution product-ion spectra and HR full-scan mass spectra, respectively. The HR mass spectra analyzes identified the analytes by using product-ion spectra along with the accurate masses with respect to the protonated molecules. Two reference materials were analyzed for segmental analysis and verification of single strand of hair specimen from real cases were accomplished (Miyaguchi and Inoue, 2011). Similarly, there are reviews that state that LC-MS and HRMS-UHPLC are useful and promising for metabolite profiling in urine samples and toxicological drugs in biological matrices (Wissenbach et al., 2011; Zhang and Watson, 2015).

6.3 ENVIRONMENTAL SAMPLE ANALYSIS (SOIL AND WATER)

The soil is a powerful source that can provide ample data of the crime scene and related evidences. The soil has chemical precipitates (salt crystals, calcium carbonate), inorganic mineral grains, organic matters, animal matters, dead plants, bacteria, insect carapace, fungi, higher plants roots and algae from soil, etc., as its constituents. Forensic investigation of soil involves comparing of its components for oxides, diatoms, organic matter, light or heavy minerals, stable radioisotopes, pollens, and microorganisms and by measuring the physical characteristics including particle size distribution, color density, etc., the soil organic matter (SOM) profile and mineralogical techniques stay harmonious to resolve forensic problem. Of these characteristics, SOM may be extensively affected by environmental factors including soil microorganism, animal residues, weathering, human beings' artifacts, plants of nearby soil. Although analytical methods like FT-IR, UV-Vis, HPLC, and pyrolysis-GC have been employed for the forensic application of SOM. However, analytical limitations pose difficulty in soil discrimination and low contents of SOM, less than 5%, compared to those of minerals. Therefore, minerals have been considered as more common soil evidence than SOM, and SOM has been barely used for soil analysis. Nevertheless, SOM can become more important in soil classification, if its components and their amounts can be exactly identified. The SOM is the largest storage of the broad organic carbon as well as carries over thrice the content of carbon from plant biomass on land. SOM has been mainly studied in geochemistry, agriculture, and the environment. Lignins, carbohydrates, cellulose, horpane, protein, lipid, or fatty acids are chemotaxonomic compounds of SOM. These constituents are emerged from microorganisms, plants, and animals. In particular, fatty acids and lignin that are found in SOM provide more information than other compounds. Lignin, existing in different tissues of gymnosperms, grasses or angiosperms. It is a three-dimensional phenolic biopolymer with regular arrays of p-hydroxyphenyl as well as guaiacyl syringyl. The contribution of lignin and its role in the global carbon cycle is significant, as it is the second most abundant natural biopolymer on earth. Because lignin exists in plants, the lignin in soil may provide information about the plants nearby. Fatty acids, originate from microorganisms and plants, and therefore provide information about the soil nearby this microorganism and plants. Beside, urea present in the animal excretion or in a fertilizer, may be considered as discriminant characteristics (Lee, 2012). Disposal of used or unwanted clothes in landfills forms another prime source in forensic investigation. Specially soil samples that contain reactive dyes from fabrics are extracted efficiently by QuEChERS method and characterized by HRMS. Dyes such as C.I. Reactive Blue 49 (RB49), C.I. Reactive Orange 35 (RO35), C.I. Reactive Blue 19 (RB19), and C.I. Reactive Black 5 (RBlk 5) are studied to find the possible degradation products in landfills. This analysis identifies traces of native dyes and their hydrolysis products in soil prone to degradation between 45 and 90 days (Sui et al., 2019).

Besides, the understanding of the major differences between human fecal contamination (sewage) and animal fecal contamination (manure) is important because the former is more harmful than latter as viruses, the basic cause of illness is born through fecal exposure (host-specific) (Field and Samadpour, 2007). Both manure and sewage are the source of nitrates that are considered by WHO as a major source of health and environmental problems, hence, it is an area of relevance for forensic studies to identify the source of contamination. There are various methods adopted to identify nitrate and their isotope contamination and to differentiate between manure and sewage like geochemical tracers, physicochemical properties of water, linking isotope and land-use type compositions and fecal indicator bacteria (Nestler et al., 2011). However, these techniques have been unsuccessful, therefore a suite of veterinary and human chemical markers such as food additives or pharmaceuticals were researched extensively that provided the greatest potential in this respect (Fenech et al., 2012). This is because of their physicochemical characteristics such as nonvolatile, water-soluble and resistant to biodegradation and has numerous co-occurring sources. Hence, varied methods have been developed for their detection for monitoring their behavior in forensics. One such efficient method (SPELCMS/MS) was developed and validated for chemical markers with a detection limit of about 50 pg/mL to differentiate manure and sewage surface water samples (Fenach et al., 2013).

To investigate water samples, especially those which have compounds with different concentrations, semi-polar, polar, complexed or undetected compounds with HRMS were used. This study utilized liquid chromatography coupled with time of flight mass spectrometer (LC-QTOF) along with passive sampling utilizing polar organic chemical integrative sampler (POCIS), which are not possible otherwise. Besides, various data processing techniques such as target (homemade database), non-target (Statistical, Principal components analysis (PCA) linked to ion intensity and original chromatograms) and suspect screening (literature database) and trend plots for identification of relevant compounds were used. This approach is not only useful for environmental pollution, but also helpful for forensic investigation (Soulier et al., 2016). A similar study was demonstrated by Gago-Ferrero et al. for about 2,316 pollutants (pesticides, doping compounds, drug of abuse, transformation products (TP), surfactants, industrial chemicals) in water by ultra-performance liquid chromatography quadrupole time-of-flight mass spectrometry

(UPLCQ-ToF-HRMS/MS) with data-independent acquisition (DIA) mode that simultaneously provides MS/MS and MS spectra. This approach incorporated detection capability (CCβ) and decision limit (CCα) to provide realistic metrics and greater successful identification rate (95%). However, this method was capable of analyzing 195 contaminants, therefore the new identification point system was introduced by a commission decision 2002/657/EC which considered mass accuracy (MA), isotopic fit, fragmentation, the RT of HRMS to detect 398 contaminants in water samples (Gago-Ferrero et al., 2019). Another approach reported that could be useful for forensic investigation was comparison of non-target analysis and post-target analysis by UPLC-TOF-MS. For both, post-target, and non-target techniques, all compounds with a mass/charge (m/z), which was within a specific mass range, eluted from the column, ionized in the ionization chamber are measured and a narrow-window plot with extracted ion chromatogram (nw-XIC) of 20–50 mDa was plotted after analysis. On the one hand, the analytes of interest corresponding to the nw-XICs of the masses are extracted from the complete datasheet, which gives a retrospective reanalysis of the sample years, after the first sample analysis, this is known as post-target approach. On the other hand, a special deconvolution software was used to detect unknown components in the sample chromatogram, which are extracted from TIC and analyzes the ions of the same unknown components. The detected elemental component from the compound was detected from the accurate ion mass, which is known as non-target techniques. However, both the approaches including post target and non-target techniques may sometimes provide false-negative results due to the incapability of analyzing all compounds in the sample. These two techniques are the inherent nature of LC-MS analysis due to the chromatographic and ionization process involved. However, post-target analysis approach seemed beneficial for analyzing contaminants in wastewater samples having psycholeptics like oxazepam, nordiazepam, and temazepam (Nurmi et al., 2012).

6.4 POISONING

Poisoning cases are frequently witnessed in forensic autopsy. However, the diagnosis of the same is difficult specially hydrogen sulfide (H_2S) poisoning because it oxidizes readily in the body. In addition, analytical methods for measuring thiosulfate such as high performance liquid chromatography (HPLC), absorption spectroscopy, and titration give poor specificity. Therefore, to improve the specificity and sensitivity as well as to simplify the method of sample preparation, studies of analyzing thiosulfate with LC-MS are initiated because it is capable of measuring ions in a sample. Jin et al. (2016) worked to develop a simple, rapid, and highly sensitive method for quantitative analysis of serum thiosulfate by LCMS. This method maintained high precision, accuracy, and sensitivity (lower quantification limit-0.5 μM) to measure the thiosulfate concentration as 57.5 μM (Jin et al., 2016). Atropine, a natural alkaloid (racemic mixture of (R)-hyoscyamine and (S)-hyoscyamine) and Scopolamine can be used either as an unintentional or intentional drug of abuse (victims' control, illegal drug dilution), contaminated food consumption or accidently ingestion of the part of the plant (especially by children) and so their identification is important in forensic toxicology. However, the basic challenge encountered during their analysis is the low concentration in biological samples. Therefore, numerous methods have been employed for their characterization of which GC-MS and LCMS are most known. Better selectivity and sensitivity is achieved by LC-MS consisting of LC-MS/MS or atmospheric pressure chemical ionization (APCI) having electrospray ionization (ESI). Depending upon the sample type, injection volume and preparation method, the LLOQs reported range between 0.02 and 10 ng/mL. In literature reported, after simple deproteinization the lowest LLOQ for rabbit plasma and human plasma was 0.05 ng/mL and 0.25 ng/mL, respectively. Hence, LCMS proves efficient in resolving poisoning cases (Koželj, 2014).

6.5 EXPLOSIVE/GUNSHOT RESIDUES (GSR)

Explosives are a treat to both individual and nation and detection of explosives is a major goal for any forensic laboratory. The explosive used for civil purposes (criminal attacks or explosive for automated teller machine (ATM)) namely ammonium nitrate fuel oil (ANFO) especially in Brazil was a major challenge which was resolved by using FT-ICR-MS with electrospray ionization (ESI (+)-FTMS) in positive and negative ion modes. The ion marker was distinctive via ESI (-)-MS. It was noticed that the ANFO ion marker was easily detected from the banknote surface which was collected from the explosion theft at ATM by using a simple, robust, and accurate single mass spectrometer with simple and moderate sonic-spray ionization mass spectrometer (EASI-MS) (Hernandes et al., 2014). Numerous terrorism attacks with homemade explosives have cost a huge on lives, especially cases in Bali (2002, 2005); Boston (2013); Brussels (2016); Madrid (2004); Manchester (2017); London (2005); Paris (2015); Oslo (2011). Therefore, it became essential to curb such situation by identifying the components of such explosives by suitable method that could trace the people behind such attacks. Therefore, both forensic and security-based investigations have great interest in such identification (Dicinoski et al., 2006; Yinon and Zitrin, 1993). Particularly, ANFO, smokeless powders (nitroglycerin (NG) and nitrocellulose), black powders (sulfur charcoal, nitrate salts), perchlorate/sugar devices or chlorate devices, which particularly leave traces of organic or inorganic residues like gunshot residues (GSR). These GSR have particular composition which are particular to the manufacturer (Johns et al., 2008; Smith et al., 1999). Mass spectrometry (MS) especially HRMS coupled with IC have recently been proved efficient for detection of ionic residues of energetic materials. They have very high resolving power of about 140,000, MA less than 5 ppm, improved method selectivity, quantification of complex matrices, increased data analysis flexibility and permits full scan acquisition of data with minimal interference of isobaricons, which have low m/z range found in low-order explosives. Also, they permit in silico recognition of new species due to retrospective, non-targeted, or suspect screening. However, such coupling form large water portions in eluents that poses a challenge to forensic analysis, which has been resolved by using organic solvent eluents like methanol or acetonitrile referred to as solvent enhanced IC. To illustrate, an approach involving ethanolic eluent for direct coupling of HRMS as well as IC with auxiliary post column infusion pumps for gas-phase transfer resulted in organic explosives integration before injection without solvent exchange. The linearity was greater than 0.99, reproducibility with the peak area (%RSDs) $< 10\%$ and mass accuracies with peak area was about ($\delta m/z$) < 3 ppm. This method was efficiently used to determine contact of swabbed hand sweat as well as latent fingermark samples with black powder substitute containing perchlorate, benzoate, and nitrate for targeted analysis. In addition, the effect of time while handling the recorded signals could be interpreted when this method works synergistically with PCA to support forensic investigations. Besides, a non-targeted application, PCA uses full scan IC-HRMS data facilitated GSR classification from three different ammunition types. Further, 15 anions along with 20 markers of GSR in silico were identified during targeted analysis. Therefore, this method provides flexibility, consistency to detect ionic energetic material for forensic analysis. Moreover, this method has wide spread application for non-targeted, targeted as well as suspect screening application in other fields by widening the separation space to low molecular weight organic and inorganic anions (Gallidabino et al., 2019). Usually after a shooting or explosive event, unburned or partially burned residues of smokeless powders were obtained. These smokeless powders are very common low grade explosives utilized as an energetic material in improvised explosive devices (IEDs) (Li et al., 2016). These samples if identified becomes prime evidence that could lead to the suspect. Commonly, Direct analysis in real-time (DART)-mass spectrometry (DART-MS), DART-time-of-flight-MS (DART-TOFMS), thermal desorption–direct analysis in real-time–high resolution mass spectrometry (TD-DART-HRMS) are used as a rapid means of analysis of the minor and the major components of such powders. Several explosives were studied via DART-MS including NG, 2,4-dinitrotoluene (2,4-DNT). NG, 2,4-DNT, and 2,6-DNT. A study compared the efficiency of DART-HRMS and TD-DART-HRMS for smokeless powders (Li et al., 2016; Mach et al., 1978; Nilles et al., 2010; Sisco et al., 2013). Besides, the samples were extracted to reduce sample variability as well as in order to permit direct and simple comparison between techniques. Numerous approaches such as Chemometric statistical approaches, Sorensen correlation, hierarchical cluster analysis (HCA), linear discriminant analysis (LDA) were applied. Sorensen correlation compared both their spectra and suggested that the negative and positive ionization mode spectra highly corresponds to each other denoting similarities between DART-HRMS and TD-DART-HRMS in the positive ionization mode spectra. HCA indicated the formation of groups in particular spectra, and LDA showcased higher classification accuracy for TD-DART-HRMS. The validation from external test indicated that TD-DART-HRMS model was more robust than DART-HRMS model. This study demonstrated the use of TD-DART-HRMS for the analysis powders that were smokeless and established their complementary nature in comparison with DART-HRMS (Lennert, 2019). Figure 6.1 provides some real examples of ANFO marker utilization.

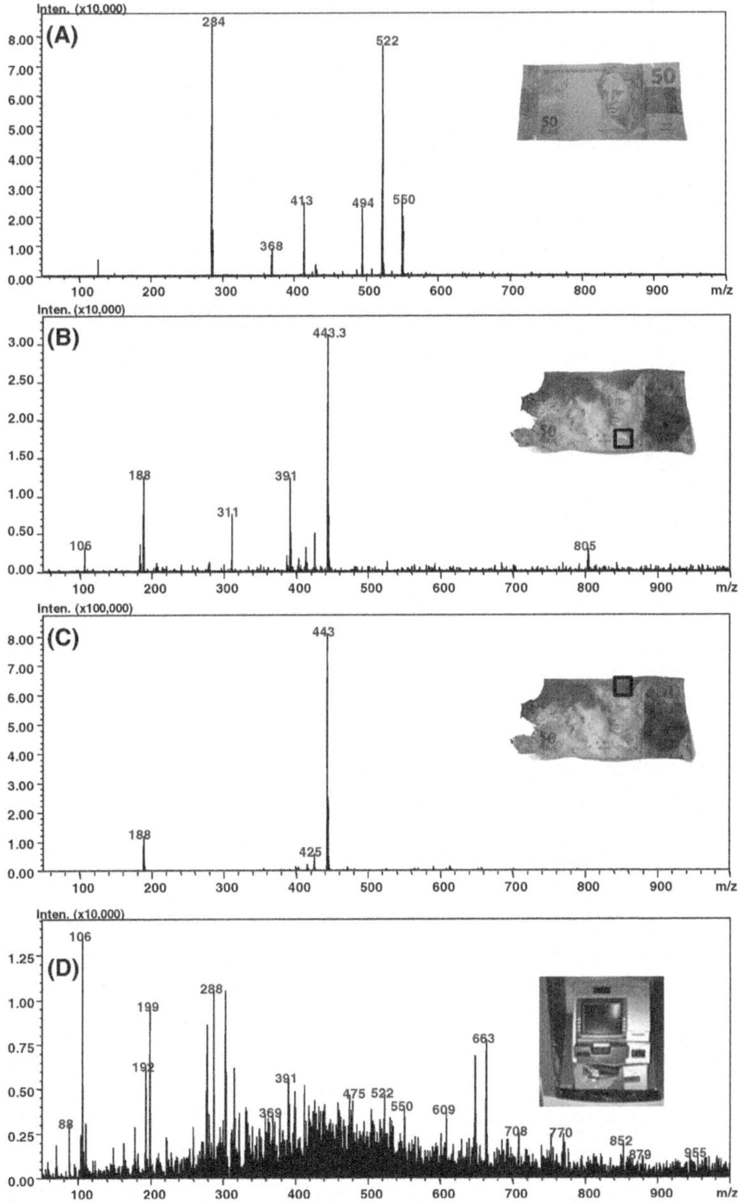

FIGURE 6.1 EASI(+)-MS from the surface of (A) an undisturbed Brazilian banknote; (B) and (C) the banknote after ANFO explosion; and (D) an extract of ANFO remaining from a real crime scene.

Source: Reprinted with permission from: Hernandes et al. (2015). © Elsevier.

6.6 INK ANALYSIS

To detect the forged content in any suspected document ink analysis is of particular interest under forensic examination as ink consists of colorants, solvents, resins, and other additives (Ezcurra et al., 2010). Many methods have been developed since past, such as ultraviolet-visible spectrophotometry, Raman spectroscopy, thin-layer chromatography (TLC), capillary electrophoresis (CE) and high performance liquid chromatography (HPLC) (Sun et al., 2016). However, forensic ink analysis focuses on the identification of dyes that are dissimilar in analyzed inks and elucidate the difference observed in chromatogram or spectra, to ameliorate the reliability as well as precision of forensic reports. Identification of dyes should provide information of the ink composition to facilitate the search for the origin of ink and other relevant details which would be useful for forensic ink library including US secret service digital ink library (Neuman et al., 2011). Further, promoting ink dating study by searching new dyes in ink that may be possible to degrade as ink ages or identifying the approximate date when the ink was manufactured or first marketed was first marketed. Under these considerations MS becomes the key technique that can be utilized with discrete ionization sources, including time of flight secondary ion mass spectrometry (TOF-SIMS), electrospray ionization MS, Direct analysis in real-time ionization-time of flight (DART-TOFMS), matrix-assisted laser desorption ionization-time of flight secondary ion mass spectrometry (MALDI-TOFMS or LDI-TOFMS), for forensic ink analysis (Sun et al., 2016). A more promising and stable technique for forensic ink analysis is liquid chromatography diode array detector-orbitrap mass spectrometry system (LC-DAD-Orbitrap MS), which is considered to a classical combination for the analysis of dyes because DAD can detect colorants after sufficient separation from LC and Orbitrap mass spectrometer provides HR, high selectivity and sensitivity. This combination is capable of measuring accurate ion masses with HR and at rapid speed, suitable for non-target identification to induce minor destruction of documents (Blanc, 2006). In a study, six triphenylmethane basic dyes (Victoria blue R, Victoria blue B, Basic blue 7, Crystal violet, Ethyl violet, Methyl violet 2 B) three sulfonic acid dyes (Acid red 52, acid blue 1 and acid blue 9) were taken for method development as reference dyes. To check the instrument performance, the nine dyes were validated and thereafter quantified, as the results of validation was stable as well as sensitive. This method was then applied as screening analysis of 20 blue ballpoint pen inks and 10 roller ball ink pens of blue color, which resulted in the detection of 10 known dyes, de-methylated product of crystal violet (TRP) and de-methyl product of acid de-methylated product of Acid blue 90, acid blue 104, acid violet 49 and Victoria blue B. In addition, the characteristic precursors of these dyes and productions were better analyzed with good MA by Orbitrap MS. These results established an appropriate method to detect and recognize potential dyes of both acidic and basic nature in blue writing inks which resulted useful in forensic investigation (Sun et al., 2016).

6.7 CONCLUSION

LCMS, HRMS, Orbitrap or their hyphenations such as TD-DART-HRMS, FT-ICR-MS provide HR, sensitivity, selectivity, and MA that form an important analysis technique for forensic investigation. It facilitates analysis of many lives threatening, abused or date-rape drugs from blood plasma or urine samples from a cadaver or putrefied bodies which is very difficult to analyze. The analyzes of soil or water samples by LCMS or HRMS are crucial because they are the paths that link the suspect and crime. Diagnosis of drugs that lead to poisoning cases are difficult to analyze, however, LCMS with high precision and accuracy measure drugs available at minimal concentration. Besides, these HR instruments with an attribute of high speed and accuracy are capable of analyzing residues from a gunshot or explosives that prove a threat to the nation. They are reported as capable of identifying ink components particularly dyes in documents that are suspected to be forged. Hence, it seems that these analyzers will dominate the affordable accurate mass instrument market in the foreseeable future and is indeed a great relief for forensic department that are able to resolve crucial, complex cases with ease and accuracy.

ACKNOWLEDGMENTS

The authors of this chapter find it an esteemed privilege to show their gratitude to the management of Aurea Biolabs (P) Ltd., Cochin, India, who were an invariable source of support to put pen to paper and encouraged us during the entire course of this chapter. We also grab this golden opportunity to express deep appreciation to our colleagues for their support that encouraged the successful completion of this task.

DECLARATION OF INTEREST

The authors profess no declaration of interest in this present chapter.

KEYWORDS

- **drug identification**
- **explosives**
- **forensic analysis**
- **high-resolution mass spectrometry**
- **ink analysis**
- **liquid chromatography-tandem mass spectrometry**
- **poisoning**

REFERENCES

Blanc, R., Espejo, T., Lo´ pez-Montes, A., Torres, D., Crovetto, G., & Navalon, A., (2006). Sampling and identification of natural dyes in historical maps and drawings by liquid chromatography with diode-array detection. *J. Chromatogr. A., 1122*, 105–113.

Broecker, S., & Pragst, F., (2012). Isomerization of cannabidiol and Δ9-tetrahydrocannabinol during positive electrospray ionization. In-source hydrogen/deuterium exchange experiments by flow injection hybrid quadrupole-time-of-flight mass spectrometry. *Rapid Commun. Mass Spectrom., 26,* 1407–1414.

Dicinoski, G. W., Shellie, R. A., & Haddad, P. R., (2006). Forensic identification of inorganic explosives by ion chromatography. *Anal. Lett., 39*, 639–657.

Dziadosz, M., Teske, J., Henning, K., Klintschar, M., & Nordmeier, F., (2018). LC-MS/MS screening strategy for cannabinoids, opiates, amphetamines, cocaine, benzodiazepines and methadone in human serum, urine, and post-mortem blood as an effective alternative to immunoassay based methods applied in forensic toxicology for preliminary examination. *Forensic Chem., 7*, 33–37.

Ezcurra, M., Gongora, J. M. G., Maguregui, I., & Alonso, R., (2010). Analytical methods for dating modern writing instrument inks on paper. *Forensic Sci. Int., 197,* 1–20.

Favretto, D., Vogliardi, S., Stocchero, G., Nalesso, A., Tucci, M., & Ferrara, S. D., (2011). High performance liquid chromatography–high resolution mass spectrometry and micropulverized extraction for the quantification of amphetamines, cocaine, opioids, benzodiazepines, antidepressants and hallucinogens in 2.5 mg hair samples. *J. Chromatograph. A, 1218* 6583–6595.

Fenech, C., Nolan, K., Rock, L., & Morrissey, A., (2013). An SPE LCeMS/MS method for the analysis of human and veterinary chemical markers within surface waters: An environmental forensics application. *Environ Pollut., 181*, 250–256.

Field, K. G., & Samadpour, M., (2007). Fecal source tracking, the indicator paradigm, and managing water quality. *Water Res., 41*, 3517–3538.

Gago-Ferrero, P., Bletsou, A. A., Damalas, D. E., Aalizadeh, R., Alygizakis, N. A., Singer, H. P., Hollender, J., & Thomaidis, N. S., (2020). Wide-scope target screening of >2000 emerging contaminants in wastewater samples with UPLC-Q-ToF-HRMS/MS and smart evaluation of its performance through the validation of 195 selected representative analytes. *J. Hazard. Mater., 387*, 121712.

Gallidabino, M. D., Irlam, R. C., Salt, M. C., O'Donnell, M., Beardah, M. S., & Barron, L. P., (2019). Targeted and non-targeted forensic profiling of black powder substitutes and gunshot residue using gradient ion chromatography e high resolution mass spectrometry (IC-HRMS). *Analytica Chimica Acta, 1072*, 1–14.

Hernandes, V. V., Franco, M. F., Santos, J. M., Melendez-Perez, J. J., Morais, D. R., Rocha, W. F. C., Borges, R., et al., (2015). Characterization of ANFO explosive by high accuracy ESI (pm)-FTMS with forensic identification on real samples by EASI (-)-MS. *Forensic Sci. Inter., 249*, 156–164.

Jang, M., Kim, J., Han, I., & Yang, W., (2015). Simultaneous determination of LSD and 2-oxo-3-hydroxy LSD in hair and urine by LC–MS/MS and its application to forensic cases. *J. Pharma. Biomed. Anal., 115*, 138–143.

Jin, S., Hyodoh, H., Matoba, K., Feng, F., Hayakawa, A., Okuda, K., Shimizu, K., et al., (2016). Development for the measurement of serum thiosulfate using LC-MS/MS in forensic diagnosis of H2S poisoning. *Legal Med., 22*, 18–22.

Johns, C., Shellie, R. A., Potter, O. G., O'Reilly, J. W., Hutchinson, J. P., Guijt, R. M., Breadmore, M. C., et al., (2008). Identification of homemade inorganic explosives by ion chromatography analysis of post-blast residues. *J. Chromatogr. A., 1182*, 205–214.

Karampela, S., Pistos, C., Moraitis, K., Stoukas, V., Papoutsis, I., Zorba, E., Koupparis, M., et al., (2015). Development and validation of a LC/MS method for the determination of Δ9-tetrahydrocannabinol and 11-carboxy-Δ9-tetrahydrocannabinol in the larvae of the blowfly *Lucilia sericata*: Forensic applications. *Sci. Justice, 55*, 472–480.

Koželj, G., Perharič, L., Stanovnik, L., & Prosen, H., (2014). Simple validated LC–MS/MS method for the determination of atropine and scopolamine in plasma for clinical and forensic toxicological purposes. *J. Pharma. Biomed. Anal., 96*, 197–206.

Lee, C. S., Sung, T. M., Kim, H. S., & Jeon, C. H., (2012). Classification of forensic soil evidences by application of THM-PyGC/MS and multivariate analysis. *J. Analytical Applied Pyrolysis., 96*, 33–42.

Lennert, E., & Bridge, C. M., (2019). Rapid screening for smokeless powders using DART-HRMS and thermal desorption DART-HRMS. *Forensic Chem., 13*, 100148.

Li, F., Tice, J., Musselman, B. D., & Hall, A. B., (2016). A method for rapid sampling and characterization of smokeless powder using sorbent-coated wire mesh and direct analysis in real-time – mass spectrometry (DART-MS). *Sci. Just., 56*, 321–328.

Lin, L., Ni, J., Lin, H., Yin, X., Qu, C., Dong, X., & Zhang, M., (2015). Types, principle, and characteristics of tandem high-resolution mass spectrometry and its applications. *RSC Adv., 5*, 107623–107636.

Maas, A., Sydow, K., Madea, B., & Hess, C., (2017). Separation of ortho, meta and para isomers of methylmethcathinone (MMC) and methylethcathinone (MEC) using LC-ESI-MS/MS: Application to forensic serum samples. *J. Chromatograph B., 1051*, 118–125.

Mach, M. H., Pallos, A., & Jones, P. F., (1978). Feasibility of gunshot residue detection via its organic constituents. Part I: Analysis of smokeless powders by combined gas chromatography – chemical ionization mass spectrometry. *J. Forensic Sci., 23*, 433–445.

Martire, K. A., Ballantyne, K. N., Bali, A., Edmond, G., Kemp, R. I., & Found, B., (2019). Forensic science evidence: Naive estimates of false-positive error rates and reliability. *Forensic Sci. Inter., 302*, 109877.

Miyaguchi, H., & Inoue, H., (2011). Determination of amphetamine-type stimulants, cocaine and ketamine in human hair by liquid chromatography/linear ion trap–orbitrap hybrid mass spectrometry. *Analyst, 136*, 3503.

Morgan, R. M., (2019). Forensic science. The importance of identity in theory and practice. *Forensic Sci. Int. Synergy*, 239–242.

Nestler, A., Berglund, M., Accoe, F., Duta, S., Xue, D., Boeckx, P., & Taylor, P., (2011). Isotopes for improved management of nitrate pollution in aqueous resources: Review of surface water field studies. *Environ. Sci. Pollut. Res., 18*, 519–533.

Neumann, C., Ramotowski, R., & Genessay, T., (2011). Forensic examination of ink by high performance think layer chromatography- The United States secret service digital ink library. *J. Chromatogr. A., 1218*, 2793–2811.

Nilles, J. M., Connell, T. R., Stokes, S. T., & Dupont, H., (2010). Durst, explosives detection using direct analysis in real-time (DART) mass spectrometry. *Propellants. Explos. Pyrotech., 35*, 446–451.

Nurmi, J., Pellinen, J., & Rantalainen, A. L., (2012). Critical evaluation of screening techniques for emerging environmental contaminants based on accurate mass measurements with time-of-flight mass spectrometry. *J. Mass Spectrom., 47*, 303–312.

Ozturk, S., Ozturk, Y. E., Yeter, O., & Alpertunga, B., (2015). Application of a validated LC–MS/MS method for JWH-073 and its metabolites in blood and urine in real forensic cases. *Forensic Sci. Inter., 257*, 165–171.

Paul, M., Ippisch, J., Herrmann, C., Guber, S., & Schultis, W., (2014). Analysis of new designer drugs and common drugs of abuse in urine by a combined targeted and untargeted LC-HR-QTOFMS approach. *Anal. Bioanal. Chem., 406*, 4425–4441.

Piotrowski, P., Bocian, S., Liwka, K. S., & Buszewski, B., (2015). Simultaneous analysis of zolpidem and its metabolite in whole blood and oral fluid samples by SPE-LC/MS for clinical and forensic purposes. *Adv. Med. Sci., 60*, 167–172.

Sisco, E., Dake, J., & Bridge, C., (2013). Screening for trace explosives by AccuTOF-DART(R): An in-depth validation study. *Forensic Sci. Int., 232*, 160–168.

Smith, K. D., McCord, B. R., MacCrehan, W. A., Mount, K., & Rowe, W. F., (1999). Detection of smokeless powder residue on pipe bombs by micellar electrokinetic capillary electrophoresis. *J. Forensic Sci., 44*, 789–794.

Song, L., Chuah, W. C., Quick, J. D., Remsen, E., & Bartmess, J. E., (2020). Nitrogen direct analysis in real time time-of-flight mass spectrometry (N2 DART-TOFMS) for rapid screening of forensic drugs. *Rapid Commun. Mass Spectrom., 34*, e8558.

Soulier, C., Coureau, C., & Togola, A., (2016). Environmental forensics in groundwater coupling passive sampling and high resolution mass spectrometry for screening. *Sci. Total Environ.,* 563, 845–854.

Sui, X., Feng, C., Chen, Y., Sultana, N., Ankeny, M., & Vinueza, N. R., (2020). Detection of reactive dyes from dyed fabrics after soil degradation via QuEChERS extraction and mass spectrometry *Anal. Methods, 12*, 179–187.

Sun, Q., Luo, Y., Yang, X., Xiang, P., & Shen, M., (2016). Detection and identification of dyes in blue writing inks by LC-DAD-orbitrap MS. *Forensic Sci. Inter., 261*, 71–81.

Wissenbach, D. K., Meyer, M. R., Remane, D., Philipp, A. A., Weber, A. A., & Maurer, H. H., (2011). Drugs of abuse screening in urine as part of a metabolite-based LC-MS n screening concept. *Anal. Bioanal. Chem., 400*, 3481–3489.

Yinon, J., & Zitrin, S., (1993). *Modern Methods and Applications in Analysis of Explosives* (pp. 66–78). John Wiley & Sons, West Sussex, UK.

Zhang, T., & Watson, D. G., (2015). A short review of applications of liquid chromatography-mass spectrometry-based metabolomics techniques to the analysis of human urine. *Analyst, 140*, 2907.

CHAPTER 7

HRMS for the Analysis of Pesticides

E. JACKCINA STOBEL CHRISTY,[1] AUGUSTINE AMALRAJ,[2] and ANITHA PIUS[1]

[1]*Department of Chemistry, The Gandhigram Rural Institute – Deemed to be University, Gandhigram, Dindigul – 624302, Tamil Nadu, India, E-mail: dranithapius@gmail.com (A. Pius)*

[2]*Aurea Biolabs (P) Ltd., R&D Center, Kolenchery, Cochin – 682311, Ernakulam, Kerala, India*

ABSTRACT

Pesticides are one of the most crucial tools used worldwide to protect agricultural products from pests, weeds, diseases, etc. However, it turns out to be a burden on the environment due to intensive use. The more the environment receive pesticide residues, the more the threat of exposing food and microorganisms to pesticides. Pesticides create a variety of acute and chronic toxicity effects in humans, animals, and microorganisms. To some extent, high resolution mass spectrometry (HRMS) has become a more outstanding commercial instrumentation used to analyze pesticide residues from different sources. This technique is very advantageous in its selectivity because it measures a compound's accurate mass even when minor structural changes to be distinguished are needed. HRMS is hugely versatile, primarily because of its capacity to allow hyphenations and hybrid instrumentations that can facilitate more conventional quantitative applications, thereby allowing highly specific qualitative as well as

High-Resolution Mass Spectrometry and Its Diverse Applications: Cutting-Edge Techniques and Instrumentation. Sreeraj Gopi, PhD, Sabu Thomas, PhD, Augustine Amalraj, PhD, & Shintu Jude, MSc (Editors)

quantitative assays. The range of applications provide a possibility for the next decade's use of mass spectrometry (MS) to analyze pesticides.

7.1 INTRODUCTION

In the field of agriculture, pesticides have contributed greatly and become an important agent for plant protection for furthering yields in agriculture to food production (Cooper and Dobson, 2007). On the other hand, pesticides are the source of many dreadful diseases causing both occupational and environmental risks. Various routes of exposure such as residues from food matrices and drinking water, can be associated with these risks. Use of more chemical pesticides for controlling pests, insects, and diseases in plants and against insect-borne diseases (malaria, dengue, encephalitis, filariasis, etc.), its exposure lead to immune suppression, birth defects, nerve damage, hormone disruption, short term hazards (causes: dizziness, headaches, eye, and skin irritations and nausea) diminished intelligence, diabetes, reproductive abnormalities, asthma, and cancer (Abhilash and Singh, 2009; Rekha et al., 2006; Kim et al., 2017; Jayaraj et al., 2016). It is not easy to define the risks related to pesticides exposure, since it involves many different factors such as period and concentration level of exposure to soil/water/plants, type of pesticide (in terms of toxicity and persistence) and the environmental peculiarities of the affected area (Arias-Estévez et al., 2008).

Pesticides residue analysis on food/water samples is essential for the protection of human health and to guarantee international trade to avoid the toxic effects on consumers and the environment. Hence, regulation agencies have set a control factor named 'maximum residue limits (MRLs) for pesticides to present in food and food materials (MacLachlan and Hamilton, 2010; Handford et al., 2015) and recently Codex 11 Alimentarius Commission developed the guidelines, international standards and codes of practice, in order to ensure the consumer health and to follow fair practices in the trading of food (Ambrus and Yang, 2016).

There are a number of pesticides used worldwide and the possibility of exposure to pesticide residues increases accordingly. It indicates the need to develop multi-residue methods (MRMs) to cover compounds of different polarities and to allow routine analysis so that the official libraries can follow the control processes in an effective mode. Thus, different selective and sensitive chromatographic-based methods have been reported to monitor pesticides in food and water samples (Loughlin et al., 2018; Choi et al., 2015; Chowdhury et al., 2013; Danezis et al., 2016). Generally in MRMs, the testing is carried out for a selected list of around 100–200 compounds of pesticide residue, which can be separated by means of liquid chromatography (LC) or gas chromatography (GC), which in turn coupled to a triple quadrupole mass spectrometer (QqQMS). But, this task is not so easy, as the analysis protocols should involve hundreds of compounds (Gómez-Ramos et al., 2013). Over the last few years, these protocols were followed, and a widely accepted target analysis workflow is resulted, where the development of acquisition method is time-consuming and extensive and resulted in the analysis of hundreds of pesticide residues simultaneously in a single run (Agüera et al., 2017). HRMS instrument present several advantages with respect to conventional QqQ systems in pesticide residue analysis and the advantages are listed in Figure 7.1.

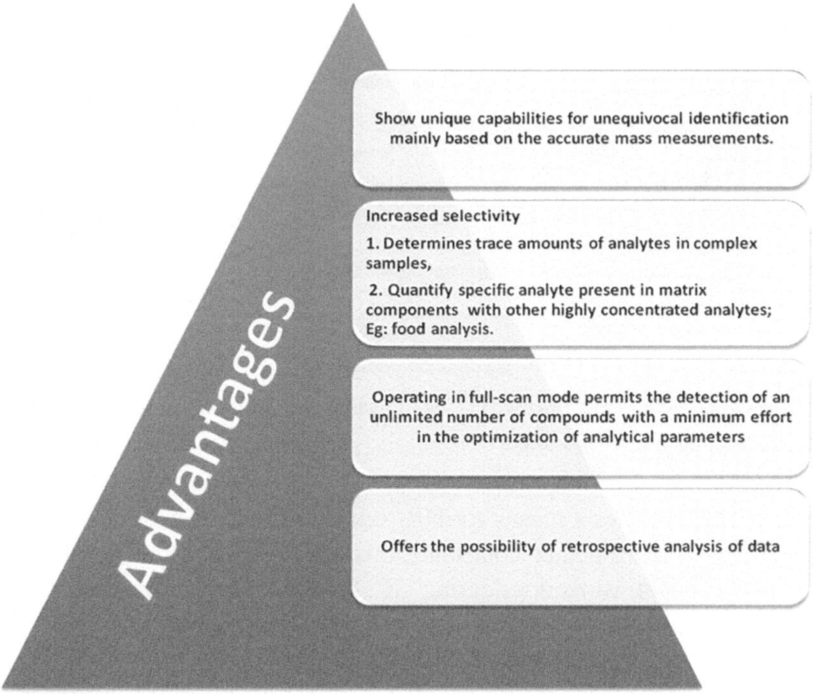

FIGURE 7.1 Advantages of HRMS over QqQ systems in pesticide residue analysis.

The interest in GC/MS applications is increasing due to its high resolution (HR) and accurate mass measurements. But the interpretation of these HR data and accurate mass measurements in a reliable manner are possible only with powerful software tools, especially during the analysis of complex matrices. The use of elemental database searching as an accurate mass library for pesticides in food using HRMS has also been described as a powerful tool for unequivocal identification. The data processing strategies of HRMS workflow are shown in Figure 7.2 (Agüera et al., 2017; Núñez et al., 2012).

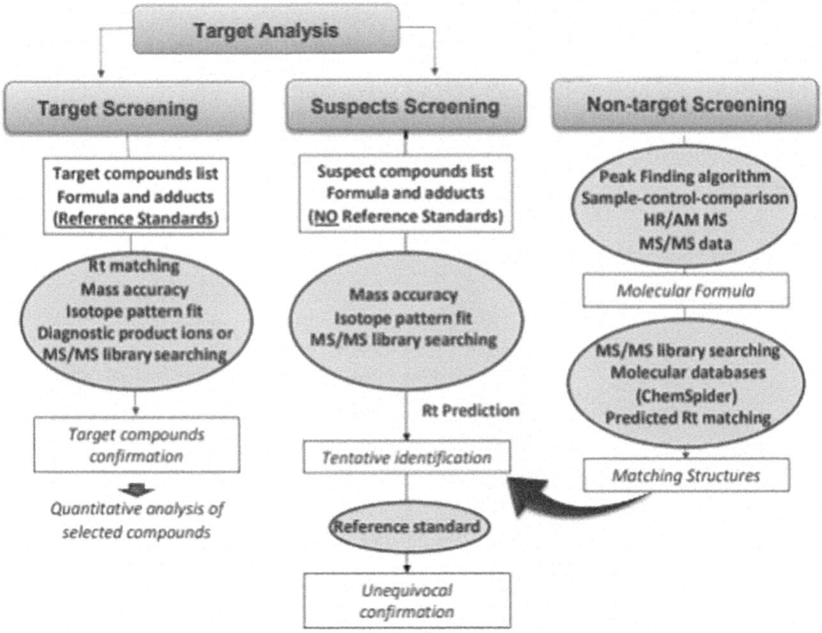

FIGURE 7.2 Main workflows used in high-resolution mass spectrometry (HRMS) methods.
Source: Reprinted with permission from Agüera et al., 2017. © Elsevier.

HRMS furnishes a means for HR, selectivity, and accurate mass determination, which together contribute towards the confirmation capability for analysis. But, while considering GC, there are only a few standardized methods available, hyphenated with HRMS, which leaves GC-MS as the major technique (Lebedev et al., 2013). HR and accurate mass refer to new generation types of instrumentation in mass spectrometry (MS), such as orbitrap, time of flight (TOF), magnetic sector, Fourier transform ion cyclotron resonance (FTICR) mass spectrometric instruments, used for full-scan MS (Ferrer et al., 2011; Kaufmann, 2012). By combining these equipments, more effective tandem analyzers can be obtained, namely hybrid ion trap/orbitrap (LTQ-Orbitrap) and quadrupole time-of-flight (Q-TOF). These tandem analyzers provide both full-scan MS and MS/MS in order to attain definite mass of precursor as well as productions for a compound identification with improved confidence and accuracy. The common ionization techniques used in these instruments are electrospray ionization (ESI), atmospheric pressure chemical ionization (APCI), matrix-assisted laser desorption ionization (MALDI), etc. (Romero-González and Garrido, 2017).

Depending on the pesticide samples, several scientists use a variety of extraction and quantification methods to estimate the pesticide residues belong to different chemical classes and present in different matrices. The extraction solvents that are used in studies pertaining to pesticide residue are selected in a way that, they should facilitate high pesticide recovery over a wide polarity range (Sharma et al., 2010). The following content focuses on different types of pesticides contaminated fruits, vegetables, and water along with the methodology adopted by various scientists to analyze their residues.

7.2 ENACTMENT OF HIGH-RESOLUTION MASS SPECTROMETRY (HRMS) IN PESTICIDE RESIDUE ANALYSIS PRESENT IN FRUITS AND VEGETABLE SAMPLES

Patel et al. developed a method for the screening of organophosphate pesticides (OPPs) (dichlorvos, methamidophos, acephate, fonofos, dimethoate, parathion methyl, ethion, phosmet, chlorpyrifos-methyl, chlorpyrifos, diazinon, pyrazophos, and etrimfos) in fruits like peach, grapes, lettuce by resistive heating gas chromatography (RH-GC) coupled with flame photometric detection (FPD) (Patel et al., 2004). Organically produced fruits extraction were made using ethyl acetate solution and analyzed on RH-GC-FPD. It took only 4.3 min for the separation of 20 pesticides. Average recoveries obtained were 70% and 116% of the spiked amount of residues at 0.01 mg/kg and 0.1 mg/kg, respectively, with relative standard deviation (RSD) of 'less than' 20%. This validated method was applied to screen around 37 pesticides from grapes and peaches. It is observed that the results obtained from the new method is in good agreement with those from MS confirmation of the same samples.

Phopin et al. (2017) extracted pesticide residues in mangosteens using the pesticide multiresidue QuEChERS (Quick Easy Cheap Effective Rugged and Safe) method and were directly injected into Bruker 456 gas chromatograph (GC) coupled with Bruker Scion Triple Quadrupole mass spectrometer (GC-MS/MS). They found that in 97% of samples, either dimethoate, chlorpyrifos, chlorothalonil, diazinon, metalaxyl or profenofos diazinon was identified to exceed their MRLs. Figure 7.3 represents levels of pesticide below and above MRL values of pesticide.

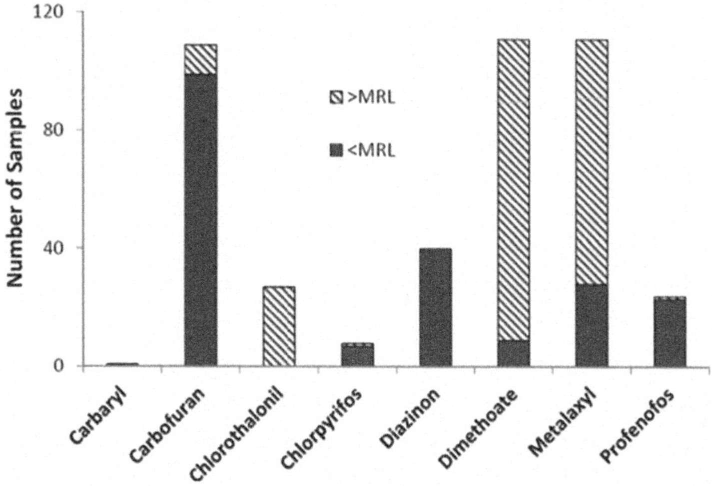

FIGURE 7.3 Types of pesticides detected in the mangosteen samples ($n = 111$). For each pesticide found, the samples were categorized according to the level of pesticide detected, i.e., pesticide detected with level of < MRL (■); and pesticide detected with level of >MRL (▧).

Source: Reprinted with permission from Agüera et al., 2017. © Elsevier.

Another Multiresidue method, proceeding with solid-phase extraction (SPE) cartridges for sample preparation was used for the analysis of 60 different pesticide compounds in fruits and vegetable samples. Detection by UHPLC/TOF-MS is accomplished by measuring the protonated molecules' accurate mass $[M+H]^+$ with a typical mass accuracy (MA) of better than two ppm. The recovery rates and RSD of the studied pesticides were 74%–111% and <13.2% respectively, for concentrations below 10 μg.kg^{-1}. Among the results, the most frequently appeared residues were Acephate, Ethion, Profenophos, and Monocrotophos which were detected in a range of 14 to 2,774, 13 to 71, 16 to 164 and 8 to 253 g.kg^{-1}, respectively. The limit of quantification obtained for the majority of compounds in the method was below the MRLs which is accepted by the Food Safety Standard Authority of India and the European Union (Sivaperumal et al., 2015).

In many developing countries, the use of pesticides has been highly encouraged to increase the production of crops per acre. A Simple and adequate method for the determination of different pesticide residues in vegetable samples was developed by Sapahin et al. by using solid-phase microextraction (SPME), which is equipped with gas chromatography-flame photometric detection (GCFPD) (Sapahin et al., 2019). The method validation was in the range of 0.1–100 µg/L with satisfied repeatabilities ranging in between 2.44% and 17.9% for all analytes. It shows a limit of detection (LOD) in the range of 0.01 to 0.14 µg/L and limit of quantitation (LOQ) in the range 0.03 to 0.42 µg/L limits of detection and quantitation respectively. The method was applied for other local vegetables like kale, cabbage, and mustard and tested samples detected with chlorpyrifos (0.22–1.68 µg/kg). But, the obtained values were lesser than the MRLs as described in the Food Act and Regulations of Malaysia. Chowdhury et al. (2013) have evaluated the presence and concentration of pesticides on vegetables grown in several areas in Bangladesh for use as a reference point for future monitoring and to allow preventive measures to be taken to minimize the human health risks. Summary of pesticide types, number of contaminated samples and levels of contamination in vegetable samples from Bangladesh are shown in Figure 7.4.

A method was successfully developed and validated for the rapid detection of 19 pesticide residues of different chemical classes in vegetable samples. López-Pérez et al. (2006) studied the dynamics of different commercially available formulations, including two insecticides, herbicides, and a combination of two fungicides. The concentrations of each active ingredient were screened over the entire growing season of the potatoes and in the 0–1 cm and 1–15 cm soil layers (where potatoes were grown). GC with mass selective detection was successfully used for the pesticide determination. Only metalaxyl pesticide was detected in potato tubers, with concentration less than half the maximum residual limit. Pan et al. (2008) analyzed OPPs residues (carbendazim, carbaryl, simazine, neonicotinoid, carbamate, and triazines present in leafy vegetables, namely spinach received from the local supermarkets in Hefei (Anhui province), China) by applying ultrasonic-assisted solvent extraction, followed by analysis in tandem mass spectrometry (LC-MS/MS). In this method, it was possible to extract six pesticide compounds in a single step by using ethyl acetate for 30 minutes, having a good recovery rate and LOQ of >83% and <1.4 g/kg respectively. Figure 7.5 depicts a LC-MS/MS total ion chromatogram of spiked spinach matrix. Moreover, the optimized method was a simple, low cost and effective strategy for the analysis of multi-residual pesticide compounds present in small levels in leafy vegetables.

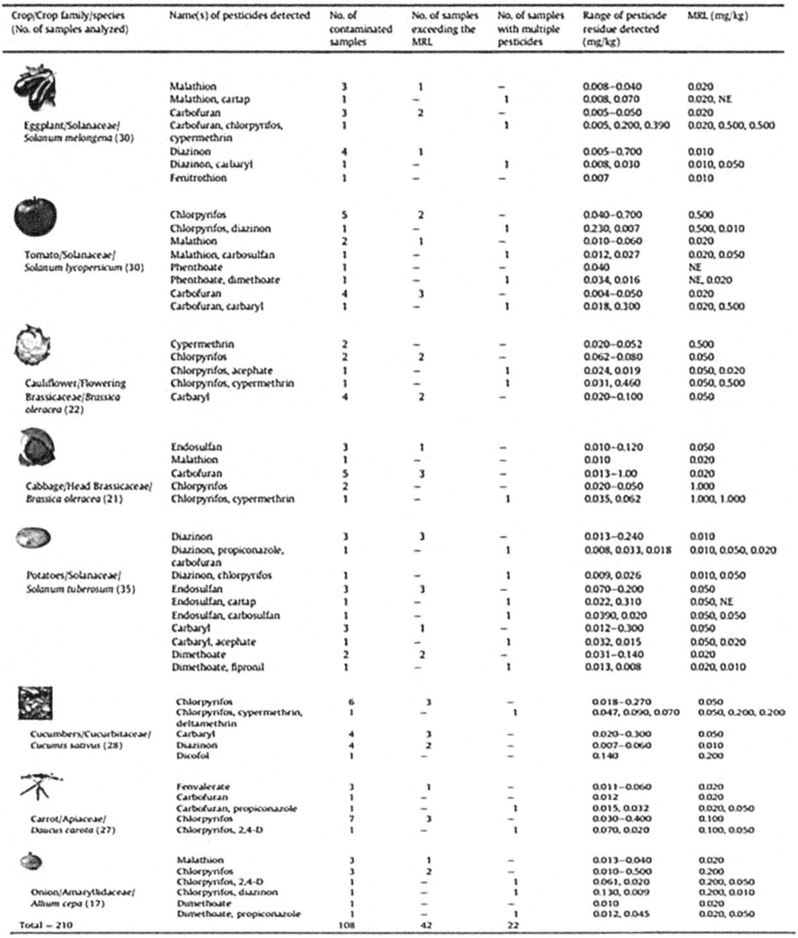

Crop/Crop family/species (No. of samples analyzed)	Name(s) of pesticides detected	No. of contaminated samples	No. of samples exceeding the MRL	No. of samples with multiple pesticides	Range of pesticide residue detected (mg/kg)	MRL (mg/kg)
Eggplant/Solanaceae/ *Solanum melongena* (30)	Malathion	3	1	–	0.008–0.040	0.020
	Malathion, cartap	1	–	1	0.008, 0.070	0.020, NE
	Carbofuran	3	2	–	0.005–0.050	0.020
	Carbofuran, chlorpyrifos, cypermethrin	1		1	0.005, 0.200, 0.390	0.020, 0.500, 0.500
	Diazinon	4	1		0.005–0.700	0.010
	Diazinon, carbaryl	1	–	1	0.008, 0.030	0.010, 0.050
	Fenitrothion	1	–	–	0.007	0.010
Tomato/Solanaceae/ *Solanum lycopersicum* (30)	Chlorpyrifos	5	2	–	0.040–0.700	0.500
	Chlorpyrifos, diazinon	1	–	1	0.230, 0.007	0.500, 0.010
	Malathion	2	1	–	0.010–0.060	0.020
	Malathion, carbosulfan	1	–	1	0.012, 0.027	0.020, 0.050
	Phenthoate	1	–	–	0.040	NE
	Phenthoate, dimethoate	1	–	1	0.034, 0.016	NE, 0.020
	Carbofuran	4	3	–	0.004–0.050	0.020
	Carbofuran, carbaryl	1	–	1	0.018, 0.300	0.020, 0.500
Cauliflower/Flowering Brassicaceae/*Brassica oleracea* (22)	Cypermethrin	2	–	–	0.020–0.052	0.500
	Chlorpyrifos	2	2	–	0.062–0.080	0.050
	Chlorpyrifos, acephate	1	–	1	0.024, 0.019	0.050, 0.020
	Chlorpyrifos, cypermethrin	1	–	1	0.031, 0.460	0.050, 0.500
	Carbaryl	4	2	–	0.020–0.100	0.050
Cabbage/Head Brassicaceae/ *Brassica oleracea* (21)	Endosulfan	3	1	–	0.010–0.120	0.050
	Malathion	1	–	–	0.010	0.020
	Carbofuran	5	3	–	0.013–1.00	0.020
	Chlorpyrifos	2	–	–	0.020–0.050	1.000
	Chlorpyrifos, cypermethrin	1	–	1	0.035, 0.062	1.000, 1.000
Potatoes/Solanaceae/ *Solanum tuberosum* (35)	Diazinon	3	3	–	0.013–0.240	0.010
	Diazinon, propiconazole, carbofuran	1	–	1	0.008, 0.033, 0.018	0.010, 0.050, 0.020
	Diazinon, chlorpyrifos	1	–	1	0.009, 0.026	0.010, 0.050
	Endosulfan	3	3	–	0.070–0.200	0.050
	Endosulfan, cartap	1	–	1	0.022, 0.310	0.050, NE
	Endosulfan, carbosulfan	1	–	1	0.0390, 0.020	0.050, 0.050
	Carbaryl	3	1	–	0.012–0.300	0.050
	Carbaryl, acephate	1	–	1	0.032, 0.015	0.050, 0.020
	Dimethoate	2	2	–	0.031–0.140	0.020
	Dimethoate, fipronil	1	–	1	0.013, 0.008	0.020, 0.010
Cucumbers/Cucurbitaceae/ *Cucumis sativus* (28)	Chlorpyrifos	6	3	–	0.018–0.270	0.050
	Chlorpyrifos, cypermethrin, deltamethrin	1	–	1	0.047, 0.090, 0.070	0.050, 0.200, 0.200
	Carbaryl	4	3	–	0.020–0.300	0.050
	Diazinon	4	2	–	0.007–0.060	0.010
	Dicofol	1	–	–	0.140	0.200
Carrot/Apiaceae/ *Daucus carota* (27)	Fenvalerate	3	1	–	0.011–0.060	0.020
	Carbofuran	1	–	–	0.012	0.020
	Carbofuran, propiconazole	1	–	1	0.015, 0.032	0.020, 0.050
	Chlorpyrifos	7	3	–	0.030–0.400	0.100
	Chlorpyrifos, 2,4-D	1	–	1	0.070, 0.020	0.100, 0.050
Onion/Amaryllidaceae/ *Allium cepa* (17)	Malathion	3	1	–	0.013–0.040	0.020
	Chlorpyrifos	3	2	–	0.010–0.500	0.200
	Chlorpyrifos, 2,4-D	1	–	1	0.061, 0.020	0.200, 0.050
	Chlorpyrifos, diazinon	1	–	1	0.130, 0.009	0.200, 0.010
	Dimethoate	1	–	–	0.010	0.020
	Dimethoate, propiconazole	1	–	1	0.012, 0.045	0.020, 0.050
Total – 210		108	42	22		

FIGURE 7.4 Summary of pesticide types, number of contaminated samples and levels of contamination in vegetable samples.

Source: Reprinted with permission from Chowdhury et al., 2013. © Elsevier.

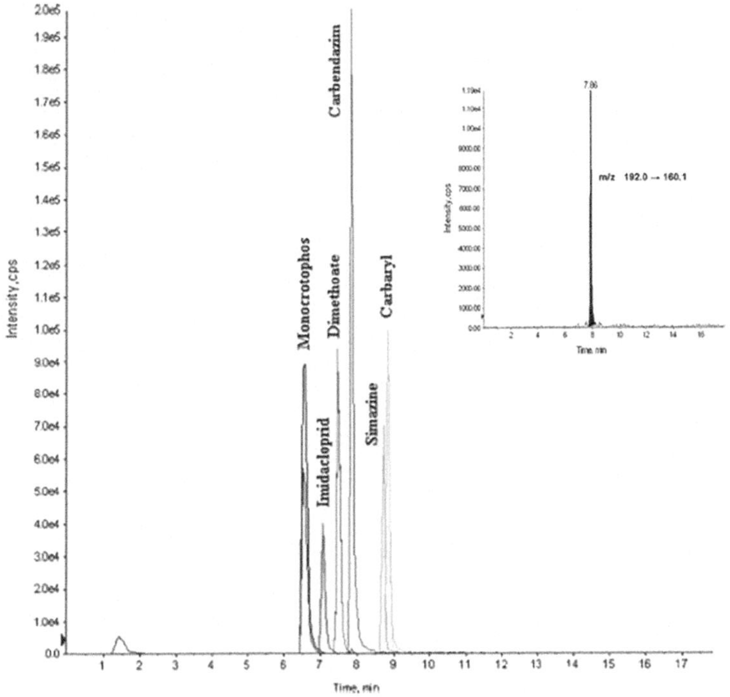

FIGURE 7.5 LC-MS-MS total ion chromatogram (MRM mode) of a spiked spinach matrix at 0.018–0.035 mg/kg after ultrasonic extraction. (Inset: Carbendazim residue in lettuce sample.).

Source: Reprinted with permission from Pan et al., 2008. © Elsevier.

González-Rodríguez et al. evaluated the presence and concentration of 12 different insecticides and 11 fungicides in leafy vegetables purchased from food stores and sellers in Ourense, NW Spain (González-Rodríguez et al., 2008). For identification of pesticides, GC-PolarisQ-ion trap-mass spectrometer was used and found that 84% (63 of 75 samples) of the total samples were contaminated with pesticide residues, and among all samples, 18 samples were reported to violate MRL value and their PTV-GC-ITMS chromatogram are shown in Figure 7.6.

Wu et al. (2017) have tested almost 19,786 vegetables samples during 2010–2014 (over four seasons) from the Xinjiang Uygur Autonomous Region (Xinjiang) of China and found 48 pesticide residues in the samples using GC-tandem mass spectrometer (GC-MS/MS) or LC-tandem mass spectrometer (LC-MS/MS). Among the samples, 6,325 (31.97%) were detected with the presence of pesticide residues, out of which, 768 samples have exceeded the MRLs. Out of the tested samples, the highest amount of pesticide residues were observed in celery, even exceeding the MRLs. Leafy vegetables and legume vegetables lists are shown in Figure 7.7. The most frequently detected pesticide was procymidone with 6.43% and some banned pesticides were also detected.

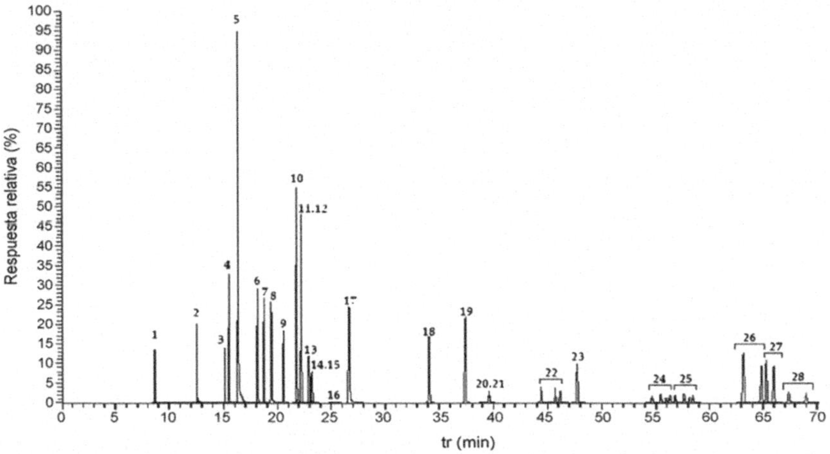

FIGURE 7.6 PTV-GC-ITMS chromatogram, SIM mode, of a pesticide standard (5 mg/L for each compound). Identification of peaks are as follow: (1) carbofuran phenol; (2) methiocarb phenol; (3) hexachlorobenzene; (4) carbofuran; (5) pyrimethanil; (6) vinclozolin; (7) metalaxyl; (8) methiocarb; (9) triadimefon; (10) cyprodinil; (11) chlorfenvinphos 1; (12) penconazole; (13) chlorfenvinphos 2; (14) triadimenol; (15) procymidone; (16) fenamiphos; (17) fludioxonil; (18) tebuconazole; (19) iprodione; (20) bifenthrin; (21) fenpropathrin; (22) lambdacyhalothrin; (23) acrinathrin; (24) cyfluthrin; (25) cypermethrin; (26) esfenvalerate; (27) tau-fluvalinate; (28) deltamethrin.

Source: Reprinted with permission from González-Rodríguez et al., 2008. © Elsevier.

Hadian et al. (2019) carried out GC-MS analysis of the vegetable samples (Cucumber, Carrot, Tomato, Cabbage, and Lettuce) for pesticide residues by using a Varian Model 3800 GC connected to an ion trap MS (ITMS) 2200 (Varian Instrument, CA, USA). The method was validated with LOD and LOQ in the range of 0.003 to 0.06 μg/g and 0.01 to 0.1 μg/g, respectively. In any of the samples, the amount of residues from the monitored pesticides were lower than the MRLs approved by FAO/WHO Codex Alimentarius Commission. Nasiri et al. (2016) have developed a method for the analysis of pesticide residues present in cucumber, which is considered to be one of the important members of the Iranian food basket. The method protocols involved modified QuEChERS sample preparation method and spiked calibration curves. Contaminated cucumber sample with diazinon, pirimicarb, and chlorpyrifos shows recovery range of pesticides at five concentration levels (n = 3) was 80.6–112.3. With an LOD of less than 10 ng/g, LOQ of less than 25 ng/g and RSD of less than 20%, the method provided a repeatable and reliable platform for determination of pesticide residues. The method was illustrated by analyzing selected pesticides in 60 garden and greenhouse cucumber samples. Results showed that 41.7% samples out of 60 samples were contaminated with pesticide residues, in which 31.7% of samples had a pesticide residue content, less than MRLs while the remaining 10% possessed higher pesticide residue content than MRLs.

	No.[a]	No.[b] (%)	No.[c] (%)
Green leaf vegetables	2909	1304 (44.83)	258 (8.87)
Legume vegetables	1139	443 (38.89)	85 (7.46)
The cabbage vegetables	2452	902 (36.79)	134 (5.46)
Gourd vegetables	2614	844 (32.29)	65 (2.49)
Solanceous fruits vegetables	4623	1489 (32.21)	103 (2.23)
Cole crops vegetables	1583	487 (30.76)	21 (1.33)
Allium vegetables	1773	490 (27.64)	92 (5.19)
Root vegetables	1903	294 (15.45)	5 (0.26)
Starchy underground vegetables	790	72 (9.11)	5 (0.63)

No.[a]: number of samples analyzed.
No.[b] (%): number of samples with pesticides detected (%).
No.[c] (%): number of samples with pesticides over the maximum residue limits (%).

FIGURE 7.7 Detection rate and over MRLs rate of vegetable groups.

Source: Reprinted with permission from Wu et al., 2017. © Elsevier.

The application of LC/TOF-MS in the determination and quantitation of pesticide residues in olive oil samples was reported by Juan Francisco García-Reyes et al. (2006). The method includes two sample treatment steps: a preliminary step based on liquid-liquid extraction and subsequent matrix solid-phase dispersion (MSPD). Then a cleanup step using florisil and acetonitrile. Accurate mass measurements (of both the protonated molecule, major fragment ion and characteristic isotope cluster) and subsequent generation of empirical formula help in the identification of compound by LC-TOFMS. The excellence of selectivity exhibited by TOF-MS can be seen in Figure 7.8. The linearity, sensitivity, MA, precision, and matrix effects are studied, and it illustrated the potential of this technique in the quantitative analyzes of herbicides in olive oil. These herbicides have an accepted MRL 100 μg/kg and the LODs of the current method lay in a better range of 1 to 5 μg/kg (García-Reyes et al., 2006).

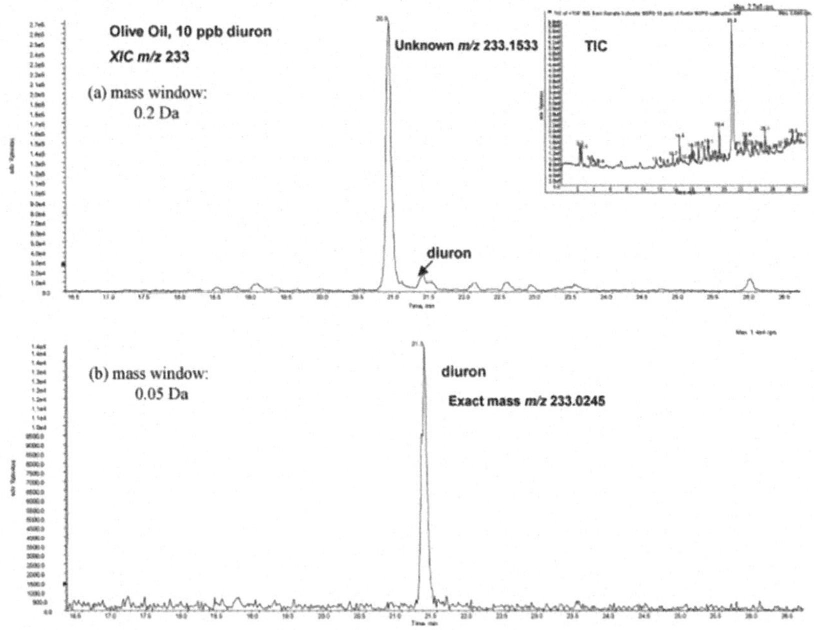

FIGURE 7.8 Extracted ion chromatogram of two different accurate mass windows: (a) 0.2; and (b) 0.05 Da for diuron at m/z 233 in an olive oil extract. Inset: TIC.

Source: Reprinted with permission from García-Reyes et al., 2006. © American Chemical Society.

7.3 PERFORMANCE OF HRMS IN THE ANALYSIS OF PESTICIDE IN GROUND/SURFACE WATER SAMPLES

Estuarine environments are being stressed continuously by novel pollution sources, derived from industry and intensive agriculture activities. Kapsi et al. (2019) have quantified pesticides (herbicides, insecticides, and fungicides) in the Louros River, North-Western Greece. Figure 7.9 shows a schematic representation of the risk assessment, environmental monitoring area and pesticide residue levels in the surface waters. A monitoring study was carried out during June 2011–May 2012, monitoring 34 compounds in the surface water from seven water sampling stations.

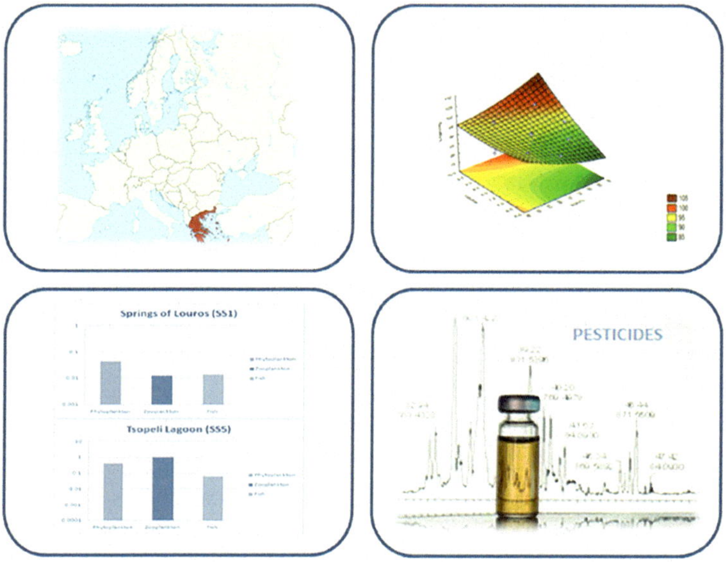

FIGURE 7.9 Schematic representation of environmental monitoring and risk assessment of pesticide residues in surface waters of the Louros River (N.W. Greece).

Source: Reprinted with permission from Kapsi et al., 2019. © Elsevier

A SPE method, coupled with GC-MS and LC-MS were developed and validated depending on the collected sample compound. The study reveals the presence of 25 pesticides including 13 herbicides, nine insecticides and three fungicides. The most predominant pesticides were quizalofop-ethyl, trifluralin, and pendimethalin. Nonetheless, tebufenpyrad was found in the highest concentration in all the sampling stations and seasons, having 0.330 μg/L at Tsopeli Lagoon exceeding the relatively low concentrations reported nationwide.

Rapid and simultaneous multi-residue method was developed for the determination of 19 metabolites and 48 pesticides in leaching water, sewage water, and tap water by using LC-MS/MS in the selected reaction monitoring (SRM). It involves an initial single-phase ultrasound facilitated extraction of samples by acetonitrile, followed by a liquid-liquid partition by NaCl solution. Recovery based on the matrix was evaluated for chosen water areas in which more than half of the compound presented very low signal suppression. The method presented good linearity in a range of 10–500 μg L^{-1} with a frequent detection limit of 0.05 ng mL^{-1}.

The recovery of the compounds, while using the method was about 74.6–111.2% with a 2.5–8.9% of RSD. This method was also used to find pesticide levels in leaching water samples collected from experimental greenhouse lysimeters in Murcia (Fenoll et al., 2011).

The presence of 29 pesticides in the agricultural fields was monitored in the Azul, Buenos Aire's southeast, San Vicente and Mista stream sub-basins. The collected samples were subjected to SPE by OASIS HLB 60 mg cartridges and UHPLC-MS/MS analysis. From the environmental monitoring results, it was found that all surface water samples were contaminated with a higher concentration of atrazine (1.4 μg L^{-1}), tebuconazole (0.035 μg L^{-1}), and diethyltoluamide (0.701 μg L^{-1}) residual pesticides based on the development of different agricultural activities (Gerónimo et al., 2014).

Xiong et al. (2019) focused mainly on Guangzhou, China's urban waterways, due to the mixing of effluents from vegetable planting and sewage, the primary contaminated zone neonicotinoid insecticides (NNIs). Modified Polar organic chemical integrative samplers (m-POCIS) was newly developed for NNIs identification and further analyzed using HPLC-MS/MS. m-POCIS for NNIs like acetamiprid, clothianidin, imidacloprid, and thiamethoxam at different concentrations (A) and water velocity (B) were shown in Figure 7.10. Results of NNIs analysis shows acetamiprid (63.5%), clothianidin (16.2%), imidacloprid (87.8%), and thiamethoxam (17.2%) with an exceeded interim chronic threshold of 35 ng/L were detected in sampling sites.

In this work, Della-Flora et al. (2019) had done a vast investigation in the hydrographic basin of Paraná State, Brazil, for pesticides using UHP-LC and UHP-GC both coupled with Q-TOF MS techniques, which helps to analyze a number of compounds, which differ in their polarity and volatility. Around 17 samples were screened in this method; thus, 52 pesticides and six metabolites were detected. Further samples were collected twice in the year to study the major transformation products (TPs) in the samples, along with the herbicide atrazine (totally 407 samples), which were analyzed using a combination of dispersive liquid-liquid microextraction (DLLME) and GC-MS measurement. The results of agrochemical atrazine concentrations in the hydrographic basin did not exceed the maximum permissible concentration of 2 μg L^{-1}, according to Brazilian legislation, and TPs desethyl atrazine and de-isopropyl atrazine maximum concentration were found to be 0.80 and 1.22 μg L^{-1}, respectively.

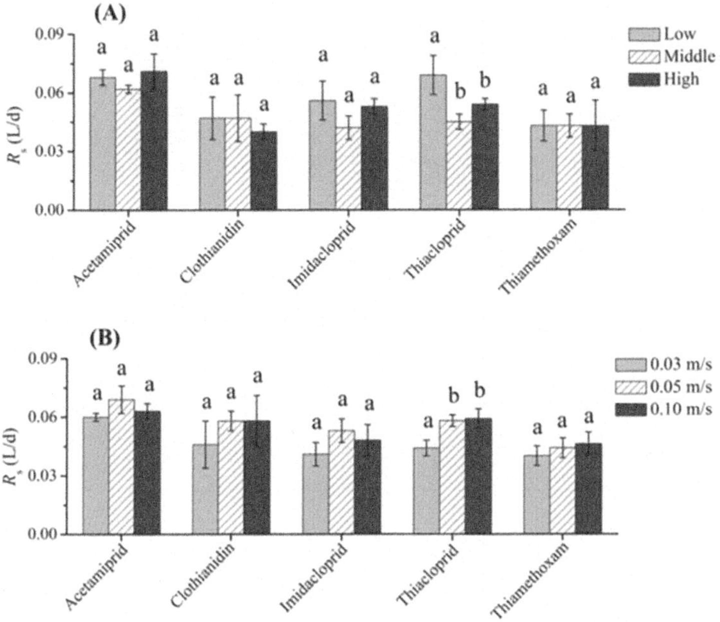

FIGURE 7.10 Calculated sampling rates (Rs) of modified polar organic chemical integrative sampler (m-POCIS) for neonicotinoid insecticides at different compound concentrations (A); and water velocity (B). Different letters indicate significant differences of Rs for each neonicotinoid.

Source: Reprinted with permission from Xiong et al., 2019. © Elsevier.

Climent et al. (2019) attempted to study surface water samples collected from the Cachapoal River basin, Central Chile. In the study, surface water samples, which were collected in the dry season were assessed for the amount of 22 pesticides and 12 degradation products in both the dissolved phase (DP) and particulate phase (PP). The analysis platform contained SPE preconcentration, followed by GC/MS, and LC/MS detection. Figure 7.11 represents the concentration and proportion of both pesticides and their degradation products in Spatio-temporal variations such as phase, site, sampling period, etc. The pesticide residue compounds present in most of the samples in the DP were atrazine, atrazine-2-hydroxy (HA), pyrimethanil, cyprodinil, and tebuconazole, whereas in the PP were HA, imidacloprid, diazinon, and pyrimidinol.

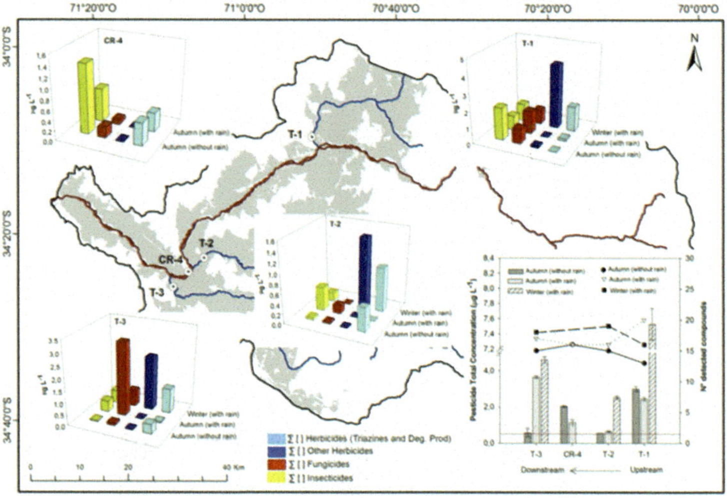

FIGURE 7.11 Total concentration of pesticides and degradation products (mg L^{-1}), number of detected compounds and spatial distribution of different pesticide groups (%) in dissolved phase of water samples collected along Cachapoal River and its tributaries. Data correspond to values determined in different sites at each period (autumn before prerain, autumn after rainfall and winter after rainfall), year 2016.

Source: Reprinted with permission from Climent et al., 2019. © Elsevier.

Herrero-Hernández et al. evaluated the existence of herbicide and insecticide residues and their seasonal variations in both the ground and surface water of the vineyard area of the La Rioja (Spain) region (Herrero-Hernández et al., 2017). The monitoring network comprises 12 surface water and 78 groundwater present in the vineyard region. Figure 7.12(1b) shows the distribution of the total amount of herbicides and insecticides (c, d) present in ground and surface water of four sampling locations. The water was examined through SPE GC-MS or by LC-MS for eight insecticides and eight major degradation products of 22 herbicides. The results revealed that, across all the campaigns, almost 65% of samples contain the herbicide named terbuthylazine and its metabolite named desethyl-terbuthylazine. The remaining compounds, like metolachlor, fluometuron, ethofumesate, and alachlor were detected only in one area, in 50% of the samples. Pirimicarb was the insecticide present in 25% of the samples, collected from March and June campaigns.

Figure 7.12(1a) shows a variation of the percentage of samples with and without detection of herbicides concentrations below or over the limit during the analyzed period. It is clear from the results that, the campaigns having highest application of these compounds have resulted with a detection of 0.5 μg L^{-1} of the same in more than half of the samples.

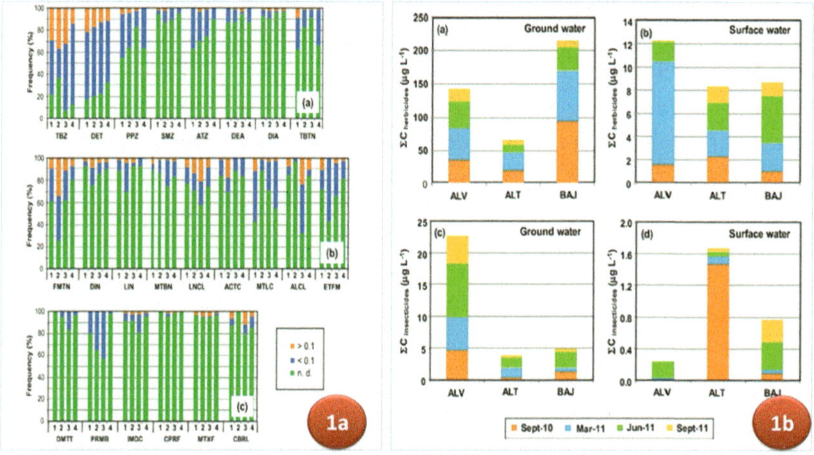

FIGURE 7.12 (1a) Variation of the percentage of samples with no detected herbicides or detected in concentrations below or over the limit established by European Directive for human consumption (0.1 μg L^{-1}) in the four sampling campaigns: Sept. -2010 (1); Mar. -11 (2); Jun. -11 (3); and Sep. -11 (4). Plots correspond to triazine herbicides (a); other herbicides (b); and insecticides (c). (1b) Distribution of the total amount of herbicides (a, b) and insecticides (c, d) in-ground and surface waters of the three subareas in the four sampling campaigns.

Source: Reprinted with permission from Herrero-Hernández et al., 2017. © Elsevier.

Quintana et al. (2019) developed a multi-residue method for *around* 100 compounds including pesticides, and some of their TP in Barcelona and its metropolitan area. An online sample extraction (0.75 mL) coupled to a fast UHPLC-MS/MS method was developed, which shows good linearity (r^2 > 0.995, with fewer residuals than 15%), accuracies, precisions, and acceptable expanded uncertainties for the monitored compounds. According to ISO/IEC 17025, the obtaining limits of quantification between 5 ng/L and 25 ng/L for all analyzed compounds. Results showed that pesticides severely contaminate the low stretch of Llobregat River and its aquifer. The maximum concentrations observed in the range of a few μg/L for DEET, propiconazole, carbendazim, diuron, and 0.1–0.5 μg/L range for bentazone, methomyl, imidacloprid, metazachlor, isoproturon, terbutryn, simazine, and tebuconazole.

7.4 MONITORING OR DETECTION OF PESTICIDES IN AIR AND SOILS USING HRMS

Guida et al. (2018) evaluated subgroups or individual chemicals of current-use pesticides (CUP) and organochlorine pesticides (OCP) in the Rio de Janeiro state. The study was prolonged for 24 months in the period of 2013–2015 in the Serra dos Órgãos National Park (SONP) and Itatiaia National Park (INP). Target pesticides were detected through a gas chromatography (GC) device coupled with mass spectrometry (GC-MS). Higher concentration of pesticide residues were observed in the atmospheric passive samples collected from SONP, when compared to the same from INP. Figure 7.13 depicts the levels of OCPs and CUPs at different seasons (2013 to 2015).

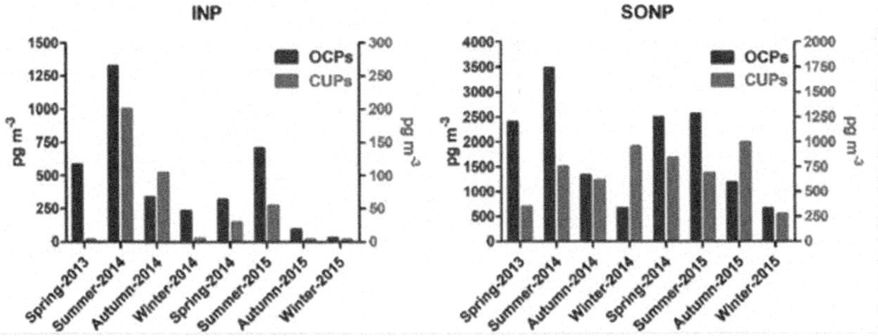

FIGURE 7.13 Seasonal variation of organochlorine pesticides (OCPs) and current-use pesticides (CUPs) air concentrations at Itatiaia (INP) and Serra dos Órgãos National Parks (SONP).

Source: Reprinted with permission from Guida et al., 2018. © Elsevier

Glinski et al. (2018) have investigated different water samples such as throughfall, surface water and stemflow in a wetland area in Georgia, USA for pesticide concentrations. More than 160 pesticide samples were screened and collectively, 32 different pesticide compounds were detected in the matrices. Most frequently and in high concentration – Metolachlor, tebuconazole, and fipronil pesticides were observed. The samples were analyzed on a LECO Pegasus® 4D GCxGC-TOF/MS equipped with a Rtx-CL Pesticides II (30 m, 0.25 mm thickness, and 0.32 mm ID; Restek, Bellefonte, PA) as the primary column and a Rxi-17 Sil MS (2 m, 0.15 mm thickness, and 0.15 mm ID; Restek Bellefonte, PA) secondary column. The predominately detected pesticides in surface water and streamflow water samples were metolachlor (0.09–10.5 mg/L), and in throughfall water samples were biphenyl (0.02–0.07 mg/L). These data help to understand the impact of indirect pesticide exposures to non-targeted habitats by analyzing both the streamflow and throughfall inputs towards surface water.

Zhang et al. (2015) have investigated the potential sources, distributions, and levels of OCPs in the environmental matrices using HRGC/HRMS technique. Pesticide extraction was carried out from 5 g of dried samples using an Accelerated Solvent Extractor (ASE300, Dionex, USA) with a mixture of n-hexane: DCM (1/1, v/v) at 150°C and at 1,500 psi (10.3 MPa), after spiking with the ^{13}C-labeled standards (OCPs-LCS). Then sample were analyzed using HRGC/HRMS system (DFS, Thermo Fisher, USA) with an electron impact (EIþ) ion source. The HRMS was operated in selected ion recording (SIR) mode at mass resolution ≥8,000. The source temperature was set to 230°C and the electron emission energy was 45 eV. In the samples from King George Island, West Antarctica, almost 23 different OCPs were detected, in different matrices. The observed levels (Σ_{23}OCPs) were in a low profile, ranging from 93.6 to 1,260 pg g^{-1} dry weight (dw) in soil and in sediment, the rage was 223–1,053 pg g^{-1} dw in moss and in lichen from 373 to 812 pg g^{-1} dw it ranged. Bar diagram shown in Figure 7.14—the given composition of HCHs (a) and DDTs (b) in soil, moss, and lichen samples. The primary contaminant in all samples was hexachlorobenzene (HCB), dichloro-diphenyl-trichloroethane (DDT) and its metabolites, primarily *p,p'*-DDE and hexachlorocyclohexanes (HCHs).

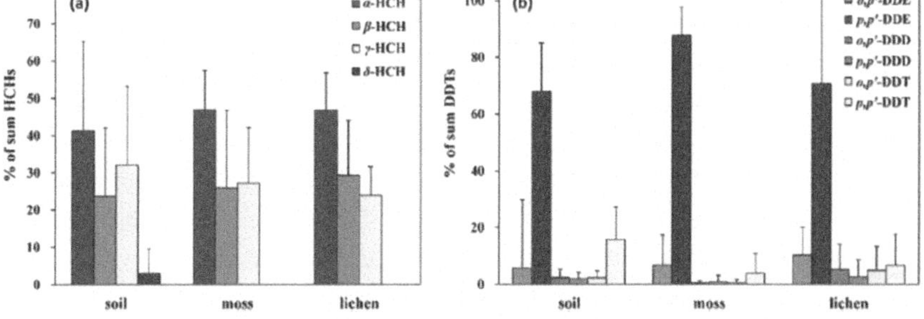

FIGURE 7.14 Composition of HCHs (a); and DDTs (b) in soil, moss, and lichen samples.

Source: Reprinted with permission from Zhang et al., 2015. © Elsevier.

Chiaia-Hernández et al. (2020) have screened 13 samples using LC-HRMS/MS and GC-MS/MS from three sediment cores of an urban and agricultural waste contaminated lake and the Swiss National Soil Monitoring Network. Soils and sediment samples were frozen-dried and subjected to pressurized liquid extraction using an in-cell cleanup technique employing sorbing agent (either Florisil or neutral alumina). An LC system coupled with QExactive™ Hybrid Quadrupole-Orbitrap Mass Spectrometer was used for the detection. GC and MS were employed in MRM mode using electron ionization (EI) for analyzing 21 compounds.

Two selective and adaptable methods were developed by Zaidona et al. (2019), in order to quantify 13 pesticide compounds, both in water and soil. An optimized and modified method was adopted using the QuEChERS extraction and a subsequent dispersive SPE (dSPE) cleanup technique and UHPLC-MS/MS for analyzing soil samples.

A 2 mm stainless steel strainer was used to homogenize and sieve surface soil samples. A sub-sample weighing 10 grams of wet weight was taken into a 50 mL polypropylene centrifuge tube and was shaken vigorously after adding 100 μL of HAc. The aliquot was spiked with 50 μL of 1,000 ng mL^{-1} of internal standard (IS) to achieve 50 ng mL^{-1} and 100 μL of 1,000 ng mL^{-1} of mixed pesticide standards to achieve 100 ng mL^{-1}. The tube was shaken for 15 seconds after adding 15 mm of MeCN. The tube was vortexed for 1 minute after adding 6 g of $MgSO_4$ and 1.5 g of NaAc. The extract was then centrifuged at 3,000 rpm for 10 minutes. A 6-ml aliquot of the supernatant was collected into a dSPE cleanup tube that was prepacked with 1,200 mg $MgSO_4$, 400 mg C18EC, and 400 mg PSA and shaken for 15 seconds. The tube was then centrifuged at 3,000 rpm for a duration of 10 minutes. The supernatant was transferred to a 15 mL polypropylene centrifuge tube and dried under nitrogen gas over a water bath at 40 °C. An one milliliter of injection solution (H_2O: MeOH, 3:1, v:v) was used to reconstitute the purified extract. By using a 0.22 μm nylon syringe filter (Membrane Solutions, USA), the extract was filtered into a 2 mL amber vial, and then proceeded for analysis in UHPLC-MS/MS analysis.

This method shows good linearity up to 500 ng mL^{-1} with an R_2 value higher than 0.9990 and 74 to 111% of recovery rate with 0.03 to 0.4 ng g^{-1} detection limit for all of the target analytes in soil. The paddy soil samples show the presence of isoprothiolane with the highest concentration (34.81 ng g^{-1}) among other quantified pesticide residues in the samples. Figure 7.15 contains the MRM chromatograms of the analytes of interest in ISs and native standards.

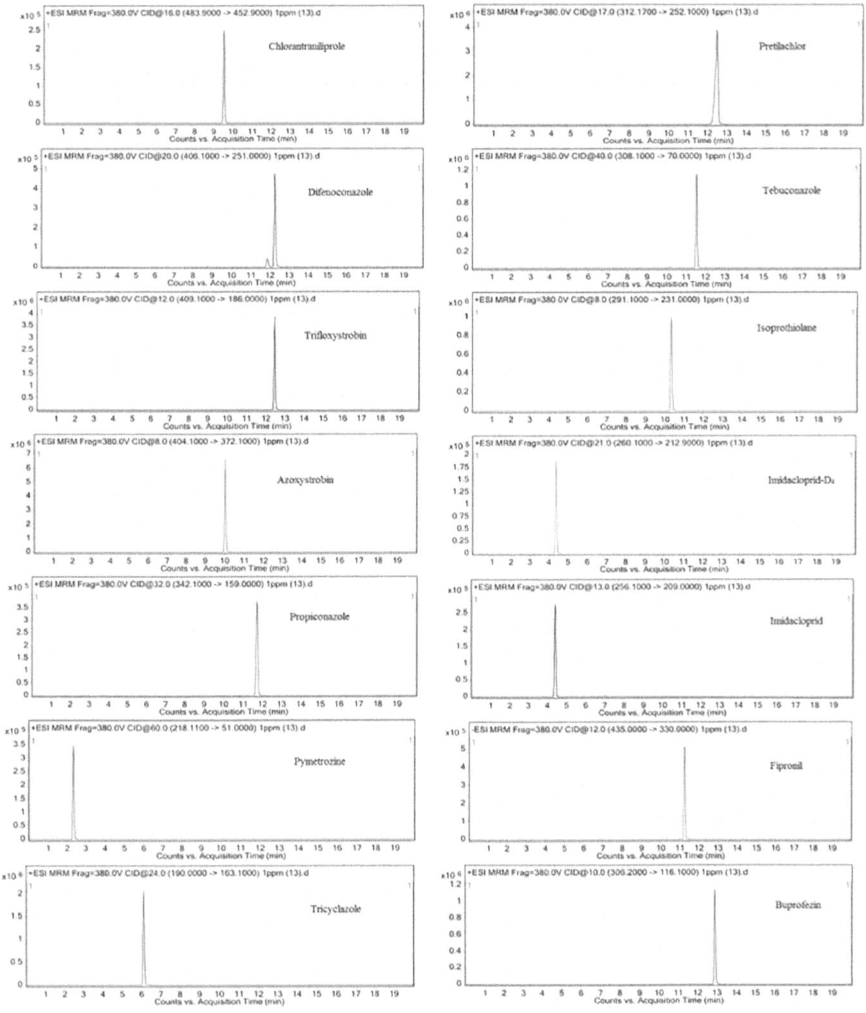

FIGURE 7.15 MRM chromatograms of the target analytes in native standards (100 ng mL^{-1}) and internal standards (50 ng mL^{-1}).

Source: Reprinted with permission from Zaidon et al., 2019. © Elsevier.

Based on the results obtained from the method's performance, developed strategies can be employed to detect pesticide residue analysis in the matrices of environmental and agricultural origin, and can manage the usage of pesticides in agricultural fields.

7.5 SUMMARY AND CONCLUSION

The capability of a HR accurate mass spectrometer in selective detection and confirmation potential of the analysis constitute the main strength of these techniques. Reliability, efficiency, sensitivity, and informational content are the primary essential features in analyzing pesticidal samples. New generation HR mass instruments provide better solution for increased resolution and MA, allowing very efficient qualitative screening methods, with a reduced number of false-positive and negative detections. Enhanced linear ranges and reproducibility have also improved quantification capabilities, and reliable initial steps are taken in this field. Altogether, it is doubtless that HRMS represents an alternative future for routine laboratories dedicated to the analysis of pesticide residues.

KEYWORDS

- **fruits**
- **gas chromatography**
- **high-resolution mass spectrometry**
- **liquid chromatography**
- **pesticides**
- **quantification**
- **time of flight**
- **vegetables**

REFERENCES

Abhilash, P. C., & Singh, N., (2009). Pesticide use and application: An Indian scenario. *J. Hazard. Mater., 165*, 1–12.

Agüera, A., Martínez-Piernas, A. B., & Campos-Mañas, M. C., (2017). Analytical strategies used in HRMS. *Appl. High Resolut. Mass Spectrom. Food Saf. Pestic. Residue Anal.*, 59–82.

Alamgir, Z. C. M., Fakhruddin, A. N. M., Nazrul, I. M., Moniruzzaman, M., Gan, S. H., & Khorshed, A. M., (2013). Detection of the residues of nineteen pesticides in fresh vegetable samples using gas chromatography-mass spectrometry. *Food Control, 34*, 457–465.

Ambrus, Á., & Yang, Y. Z., (2016). Global harmonization of maximum residue limits for pesticides. *J. Agric. Food Chem., 64,* 30–35.

Arias-Estévez, M., López-Periago, E., Martínez-Carballo, E., Simal-Gándara, J., Mejuto, J. C., & García-Río, L., (2008). The mobility and degradation of pesticides in soils and the pollution of groundwater resources. *Agric. Ecosyst. Environ., 123,* 247–260.

Chiaia-Hernández, A. C., Scheringer, M., Müller, A., Stieger, G., Wächter, D., Keller, A., Pintado-Herrera, M. G., et al., (2020). Target and suspect screening analysis reveals persistent emerging organic contaminants in soils and sediments. *Sci. Total Environ., 740,* 140181.

Choi, S., Kim, S., Shin, J. Y., Kim, M., & Kim, J. H., (2015). Development and verification for analysis of pesticides in eggs and egg products using QuEChERS and LC-MS/MS. *Food Chem., 173,* 1236–1242.

Climent, M. J., Herrero-Hernández, E., Sánchez-Martín, M. J., Rodríguez-Cruz, M. S., Pedreros, P., & Urrutia, R., (2019). Residues of pesticides and some metabolites in dissolved and particulate phase in surface stream water of Cachapoal River basin, central Chile. *Environ. Pollut., 251,* 90–101.

Cooper, J., & Dobson, H., (2007). The benefits of pesticides to mankind and the environment. *Crop Prot., 26,* 1337–1348.

Danezis, G. P., Anagnostopoulos, C. J., Liapis, K., & Koupparis, M. A., (2016). Multi-residue analysis of pesticides, plant hormones, veterinary drugs and mycotoxins using HILIC chromatography – MS/MS in various food matrices. *Anal. Chim. Acta, 942,* 121–138.

De Gerónimo, E., Aparicio, V. C., Bárbaro, S., Portocarrero, R., Jaime, S., & Costa, J. L., (2014). Presence of pesticides in surface water from four sub-basins in Argentina. *Chemosphere, 107,* 423–431.

De Souza, G. Y., Meire, R. O., Torres, J. P. M., & Malm, O., (2018). Air contamination by legacy and current-use pesticides in Brazilian mountains: An overview of national regulations by monitoring pollutant presence in pristine areas. *Environ. Pollut., 242,* 19–30.

Della-Flora, A., Wielens, B. R., Frederigi, B. S., Theodoro, T. A., Cordeiro, G. A., Ibáñez, M., Portolés, T., et al., (2019). Comprehensive investigation of pesticides in Brazilian surface water by high resolution mass spectrometry screening and gas chromatography-mass spectrometry quantitative analysis. *Sci. Total Environ., 669,* 248–257.

Fenoll, J., Hellín, P., Martínez, C. M., Flores, P., & Navarro, S., (2011). Determination of 48 pesticides and their main metabolites in water samples by employing sonication and liquid chromatography-tandem mass spectrometry. *Talanta., 85,* 975–982.

Ferrer, I., Thurman, E. M., & Zweigenbaum, J., (2011). LC/TOF-MS analysis of pesticides in fruits and vegetables: The emerging role of accurate mass in the unambiguous identification of pesticides in food. *Methods Mol. Biol., 747,* 193–218.

García-Reyes, J. F., Ferrer, C., Thurman, E. M., Fernández-Alba, A. R., & Ferrer, I., (2006). Analysis of herbicides in olive oil by liquid chromatography time-of-flight mass spectrometry. *J. Agric. Food Chem., 54,* 6493–6500.

Glinski, D. A., Purucker, S. T., Van, M. R. J., Black, M. C., & Henderson, W. M., (2018). Analysis of pesticides in surface water, stemflow, and throughfall in an agricultural area in South Georgia, USA. *Chemosphere, 209,* 496–507.

Gómez-Ramos, M. M., Ferrer, C., Malato, O., Agüera, A., & Fernández-Alba, A. R., (2013). Liquid chromatography-high-resolution mass spectrometry for pesticide residue analysis in fruit and vegetables: Screening and quantitative studies. *J. Chromatogr. A., 1287,* 24–37.

González-Rodríguez, R. M., Rial-Otero, R., Cancho-Grande, B., & Simal-Gándara, J., (2008). Determination of 23 pesticide residues in leafy vegetables using gas chromatography-ion trap mass spectrometry and analyte protectants. *J. Chromatogr. A., 1196, 1197,* 100–109.

Hadian, Z., Eslamizad, S., & Yazdanpanah, H., (2019). Pesticide residues analysis in Iranian fruits and vegetables by gas chromatography-mass spectrometry. *Iran. J. Pharm. Res., 18,* 275–285.

Handford, C. E., Elliott, C. T., & Campbell, K., (2015). A review of the global pesticide legislation and the scale of challenge in reaching the global harmonization of food safety standards. *Integr. Environ. Assess. Manag., 11,* 525–536.

Herrero-Hernández, E., Rodríguez-Cruz, M. S., Pose-Juan, E., Sánchez-González, S., Andrades, M. S., & Sánchez-Martín, M. J., (2017). (Spain). *Sci. Total Environ., 609,* 161–171.

Jayaraj, R., Megha, P., & Sreedev, P., (2016). Organochlorine pesticides, their toxic effects on living organisms and their fate in the environment. *Interdiscip. Toxicol., 9,* 90–100.

Kapsi, M., Tsoutsi, C., Paschalidou, A., & Albanis, T., (2019). Environmental monitoring and risk assessment of pesticide residues in surface waters of the Louros River (N.W. Greece). *Sci. Total Environ., 650,* 2188–2198.

Kaufmann, A., (2012). *High Mass Resolution Versus MS/MS* (1st edn.). Elsevier: B.V.

Kim, K. H., Kabir, E., & Jahan, S. A., (2017). Exposure to pesticides and the associated human health effects. *Sci. Total Environ., 575,* 525–535.

Lebedev, A. T., Polyakova, O. V., Mazur, D. M., & Artaev, V. B., (2013). The benefits of high resolution mass spectrometry in environmental analysis. *Analyst, 138,* 6946–6953.

López-Pérez, G. C., Arias-Estévez, M., López-Periago, E., Soto-González, B., Cancho-Grande, B., & Simal-Gándara, J., (2006). Dynamics of pesticides in potato crops. *J. Agric. Food Chem., 54,* 1797–1803.

Mac Loughlin, T. M., Peluso, M. L., Etchegoyen, M. A., Alonso, L. L., De Castro, M. C., Percudani, M. C., & Marino, D. J. G., (2018). Pesticide residues in fruits and vegetables of the Argentine domestic market: Occurrence and quality. *Food Control., 93,* 129–138.

MacLachlan, D. J., & Hamilton, D., (2010). Estimation methods for maximum residue limits for pesticides. *Regul. Toxicol. Pharmacol., 58,* 208–218.

Nasiri, A., Amirahmadi, M., Mousavi, Z., Shoeibi, S., Khajeamiri, A., & Kobarfard, F., (2016). A multi-residue GC-MS method for determination of 12 pesticides in cucumber. *Iran. J. Pharm. Res., 15,* 809–816.

Núñez, O., Gallart-Ayala, H., Ferrer, I., Moyano, E., & Galceran, M. T., (2012). Strategies for the multi-residue analysis of 100 pesticides by liquid chromatography-triple quadrupole mass spectrometry. *J. Chromatogr. A., 1249,* 164–180.

Pan, J., Xia, X. X., & Liang, J., (2008). Analysis of pesticide multi-residues in leafy vegetables by ultrasonic solvent extraction and liquid chromatography-tandem mass spectrometry. *Ultrason. Sonochem., 15* 25–32.

Patel, K., Fussell, R. J., MacArthur, R., Goodall, D. M., & Keely, B. J., (2004). Method validation of resistive heating - gas chromatography with flame photometric detection for the rapid screening of organophosphorus pesticides in fruit and vegetables. *J. Chromatogr. A., 1046,* 225–234.

Phopin, K., Wanwimolruk, S., & Prachayasittikul, V., (2017). Food safety in Thailand. 3: Pesticide residues detected in mangosteen (*Garcinia mangostana* L.), queen of fruits. *J. Sci. Food Agric., 97,* 832–840.

Quintana, J., De La Cal, A., & Boleda, M. R., (2019). Monitoring the complex occurrence of pesticides in the Llobregat basin, natural and drinking waters in Barcelona metropolitan area (Catalonia, NE Spain) by a validated multi-residue online analytical method. *Sci. Total Environ., 692,* 952–965.

Rekha, Naik, S. N., & Prasad, R., (2006). Pesticide residue in organic and conventional food-risk analysis. *J. Chem. Heal. Saf., 13,* 12–19.

Romero-González, R., & Garrido, F. A., (2017). Application of HRMS in pesticide residue analysis in food from animal origin. *Appl. High Resolut. Mass Spectrom. Food Saf. Pestic. Residue Anal.,* 203–232.

Sapahin, H. A., Makahleh, A., & Saad, B., (2019). Determination of organophosphorus pesticide residues in vegetables using solid-phase micro-extraction coupled with gas chromatography–flame photometric detector. *Arab. J. Chem., 12,* 1934–1944.

Sharma, D., Nagpal, A., Pakade, Y. B., & Katnoria, J. K., (2010). Analytical methods for estimation of organophosphorus pesticide residues in fruits and vegetables: A review. *Talanta, 82,* 1077–1089.

Sivaperumal, P., Anand, P., & Riddhi, L., (2015). Rapid determination of pesticide residues in fruits and vegetables, using ultra-high-performance liquid chromatography/time-of-flight mass spectrometry. *Food Chem., 168,* 356–365.

Wu, L., Zhou, X., Zhao, D., Feng, T., Zhou, J., Sun, T., Wang, J., & Wang, C., (2017). Seasonal variation and exposure risk assessment of pesticide residues in vegetables from Xinjiang uygur autonomous region of China during 2010–2014. *J. Food Compos. Anal., 58,* 1–9.

Xiong, J., Wang, Z., Ma, X., Li, H., & You, J., (2019). Occurrence and risk of neonicotinoid insecticides in surface water in a rapidly developing region: Application of polar organic chemical integrative samplers. *Sci. Total Environ., 648,* 1305–1312.

Zaidon, S. Z., Ho, Y. B., Hamsan, H., Hashim, Z., Saari, N., & Praveena, S. M., (2019). Improved QuEChERS and solid-phase extraction for multi-residue analysis of pesticides in paddy soil and water using ultra-high-performance liquid chromatography-tandem mass spectrometry. *Microchem. J., 145,* 614–621.

Zhang, Q., Chen, Z., Li, Y., Wang, P., Zhu, C., Gao, G., Xiao, K., et al., (2015). Occurrence of organochlorine pesticides in the environmental matrices from King George Island, West Antarctica. *Environ. Pollut., 206,* 142–149.

CHAPTER 8

Applications of HRMS in Environmental Impact Assessment

S. SUBASINI,[1] E. JACKCINA STOBEL CHRISTY,[1] AUGUSTINE AMALRAJ,[2] and ANITHA PIUS[1]

[1]Department of Chemistry, The Gandhigram Rural Institute – Deemed to be University, Gandhigram, Dindigul – 624302, Tamil Nadu, India, E-mail: dranithapius@gmail.com (A. Pius)

[2]Aurea Biolabs (P) Ltd., R&D Center, Kolenchery, Cochin – 682311, Ernakulam, Kerala, India

ABSTRACT

The high-resolution mass spectrometry (HRMS) technique is considered the most powerful analytical technique, which equips highly sensitive detection, identification, and unerring interpret of chemical components present in the sample, in small amounts. The benefits of HRMS are outstanding and widely grasped in various fields. However, regarding environmental chemistry, HRMS is very infrequent. The potential of mass analyzers has increased dramatically, by their coupling with ambient ionization techniques. These techniques allow simple, direct, speedy, and non-destructive ionization sources make HRMS to play a vital role in the characterization of chemical evidence collected from a crime scene like explosion or fire scenes, etc., and subjects seized for illegal practices. As environmental problems become more critical in contemporary society

High-Resolution Mass Spectrometry and Its Diverse Applications: Cutting-Edge Techniques and Instrumentation. Sreeraj Gopi, PhD, Sabu Thomas, PhD, Augustine Amalraj, PhD, & Shintu Jude, MSc (Editors)

and sophisticated HRMS are available, it is rational to expect an upgrade in accuracy, reliability, and quality of the information provided by HRMS analyses.

8.1 INTRODUCTION

The utility of HRMS is realized chiefly in various applications. Earlier, HRMS was not much used for the environmental samples because of the assumption that low-resolution mass spectrometry (MS) gives ample and authentic measurements in most cases dealing with pollutants (Lebedev et al., 2013). HRMS can perceive analytes atomic mass units to the extent of 0.001 decimals. It runs based on various techniques like Fouricr transform ion cyclotron resonance (FTICR), time of flight (TOF) and orbitrap, which are the types of HRMS used in recent times. Orbitrap technology is an ion trapping mechanism that radially traps ions about a central spindle electrode. The inner spindle-like electrode is coaxial with the outer electrode of barrel shape. The mass of analyte is measured from the frequency of harmonic ion oscillations along the axis of the electric field (Hu et al., 2005). Usually, the accurate mass of analytes can be measured even without fragmentation; in some cases, fragmentation is also possible. Fragmentation and measurements can be achieved with a quadrupole, which adds more selectivity in measurement (Wu et al., 2012). What makes the technique so distinct is that it can measure the accurate mass (exact mass) of the analyte; even minor changes in structure can be distinguished (Cook-Botelho et al., 2017).

For the past 20 years, MS has gained much contemplation in qualitative applications, detection, and endorsement of organic contaminants, investigation of metabolites, elucidation of unknowns' structure by coupling with different techniques such as liquid chromatography (LC), gas chromatography (GC), capillary electrophoresis (CE), supercritical fluid chromatography (SFC), etc., during the last decade, there was a boom in the applications of orbitrap mass analyzers, TOF, and FTICR. HRMS has started its applications in the detection of organic contaminants, antibiotics, drugs of abuse, pesticides, pharmaceuticals, for a screening of vast numbers of pollutants from industries. HRMS became superior, attributing to its performance towards both targeted and non-targeted analysis, pre-and post-target research, retrospective analysis and the characterization of transformation and metabolite products with sensitivity and accuracy. All these tasks mentioned above are pertinent to environmental sciences. So, HRMS has a wide-scope for environmental samples, one of the leading applications (Hernández et al., 2012). This chapter covers the ongoing use of HRMS in environmental sciences, and delivers a detailed picture of the standard procedures following for each analytical method.

8.2 HRMS IN THE ANALYSIS OF ANTIBIOTICS

Nowadays, human and veterinary pharmaceuticals are used in large quantities in everyday life, which results in unregulated contaminants in the form of antibiotics in an environment that becomes a problem. The widespread concern about this kind of contamination is reflected in the enormous reviews published during the last half-decade. In some recent studies, MS technique has been successfully used for the determination of antibiotics. The reliable determination of antibiotics is difficult because only a low concentration of antibiotics was perceived in the sample; moreover, each antibiotic has diverse physicochemical characteristics. Thus, compassionate, selective methods are required for monitoring antibiotics in the analytes.

Ibánez et al. (2009) have used a method incorporating TOFMS, hyphenated with ultra-high pressure liquid chromatography, for monitoring the presence of antibiotics in water samples. Water samples of different types (surface and wastewaters) were collected in polystyrene bottles (1 L) from July to November 2005 in selected sites from a Spanish Mediterranean area (Valencia region). Samples were immediately stored at <–18°C until analysis. They were thawed at room temperature, prior to analysis. Initial sampling point (4 samples) consist of sewage hospital water (SHW). The second and third sampling points were the influent (3 samples) and effluent (4 samples) of a treatment plant, receiving this SHW. Surface water from the Millars River at 0.5 km and 2.5 km downstream from the effluent discharging point (final sampling point) was also collected (4 samples). As the TOFMS untargeted analysis protocol allow the full spectrum of measurement, which in turn provides the possibility for detecting any other contaminants in the samples. This pathway have permitted the occurrence and confirmation of codeine, paracetamol, and omeprazole, among others. UPLC-QTOF-MS has been shown as a desirable technique for the detection and confirmation of pharmaceuticals in water with minimal sample handling. In their work, 42 antibiotics from 9 different families were studied.

Ibánez et al. (2009) prepared individual stock solutions at around 500 mg/L by dissolving 50 mg of the stable reference standard in 100 mL of MeOH for quinolones and CAN-water (75:25) for cephalosporins and penicillins. Other compounds were dissolved in ACN. The stock solutions were stored in a freezer at –20°C and prepared annually. Intermediate solutions were prepared in acetonitrile and were also stored at –20°C. An out-to-out intermediate mix solution was prepared weekly in acetonitrile at a concentration of 50 g/L. Working solutions were prepared daily.

Figure 8.1 shows the exact ion chromatogram for several antibiotics observed in a sample of effluent wastewater. Figure 8.2 illustrates the confirmation of clarithromycin and trimethoprim by QTOF in a raw hospital wastewater sample. The accurate masses and relative intensities of the sample's primary ions were compared with their theoretical values.

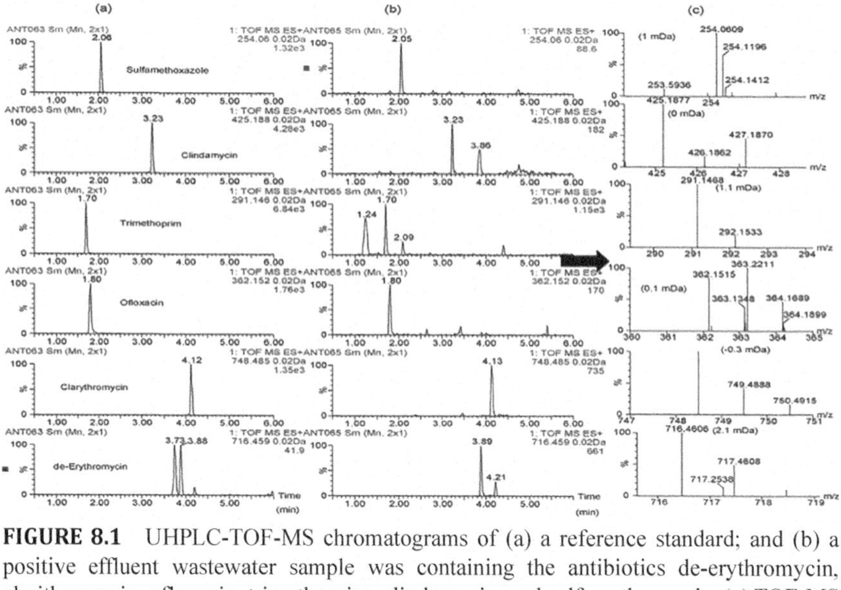

FIGURE 8.1 UHPLC-TOF-MS chromatograms of (a) a reference standard; and (b) a positive effluent wastewater sample was containing the antibiotics de-erythromycin, clarithromycin, ofloxacin, trimethoprim, clindamycin, and sulfamethoxazole; (c) TOF-MS spectra of the real sample.

Source: Reprinted with permission from Ibanez et al., 2009. © Elsevier.

Lee et al. (2000) composted livestock manures in solid and liquid form, used as fertilizers in cropland. Water samples were collected from Anseong-Si city and Hongseong-gun located at the south of Gyeonggi-do, ROK. Collected samples were centrifuged at 4°C to eliminate suspended solids. After filtering the supernatant of the centrifuged sample through a 0.2 μm polyvinylidene fluoride syringe filter. The pH of each sample was adjusted to 6.5 using a 1.5 M citrate buffer (0.5 mL per 50 mL water sample), then added 0.5 mL of 5% EDTA solution to the 50 mL of water samples. The extraction and analysis of the samples were completed using an online solid-phase extraction (SPE)-UHPLC-q-orbitrap HRMS hyphenation within 7 days.

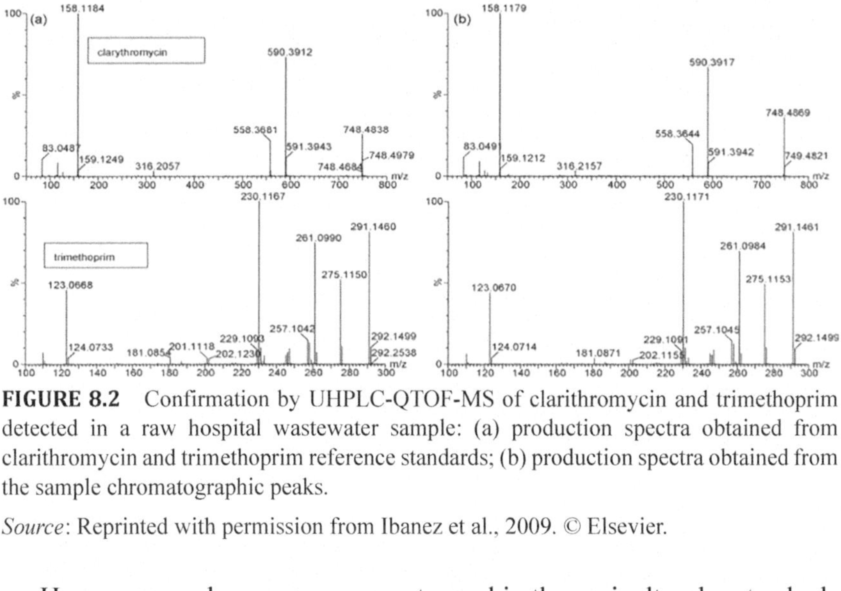

FIGURE 8.2 Confirmation by UHPLC-QTOF-MS of clarithromycin and trimethoprim detected in a raw hospital wastewater sample: (a) production spectra obtained from clarithromycin and trimethoprim reference standards; (b) production spectra obtained from the sample chromatographic peaks.

Source: Reprinted with permission from Ibanez et al., 2009. © Elsevier.

However, surplus manure amounts used in the agricultural watersheds furnishes nonpoint pollution origins as they contain higher amount of nutrients and heavy metals along with other antibiotics, including veterinary antibiotics (VAs). UHPLC-q-orbitrap HRMS equipped with an online SPE was used for the simultaneous analysis of 21 VAs belonging to nine different classes happen to find in agricultural watersheds. The quantitative amount of VAs in stream water was measured and compared according to antibiotic class for the actual environmental samples.

Figure 8.3 shows the concentrations of the VAs detected in the Gwang-cheon stream and Cheongmi watershed during the four seasons in 2019 at different sampling points. Different colors in the graph distinguish each antibiotic class, and different chemicals are represented using different patterns. The use of a hybrid quadrupole time-of-flight mass analyzer permits the authentic justification of antibiotics detected in samples, as the obtained result contained a complete full scan production spectrum, with accurate mass.

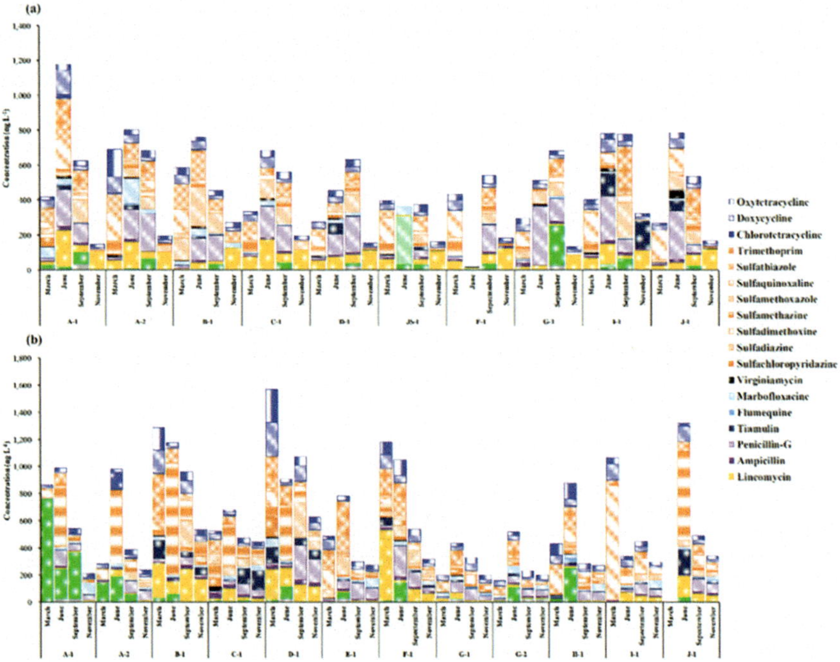

FIGURE 8.3 The residual cumulative concentration of veterinary antibiotics selected herein by sampling points and sampling time in: (a) Cheongmi stream at Anseong-Si; and (b) Gwangcheon stream at Hongseong-gun; gray: cephem, green: ionophore, yellow: lincosamide, purple: penicillin, indigo: pleuromutilin, cyan: quinolone, black: streptogramin, orange: sulfonamide, blue: tetracycline.

Source: Reprinted with permission from Lee et al., 2020. © Elsevier.

8.3 APPLICATION OF HRMS IN IDENTIFYING HYPNOTIC DRUGS

Hypnotic drugs are defined as the chemicals, which affect the body, especially the brain with long lasting and permanent consequences. Different drugs can possess different effects, and it depends upon the drugs used and delivered (including injection, inhalation, and ingestion). Presently, over 7 million people suffer from an illicit drug addiction, which causes more deaths, illnesses, disabilities, and can cause a higher risk for unintentional accidents, injuries, and domestic violence incidents (from https://www.gatewayfoundation.org/faqs/effects-of-drug-abuse/).

HRMS is helpful in detecting the level or presence of medicines and illicit drug content in biological fluids and tissues, serum, skin, urine, blood, fingerprints, hair, etc., due to its potential in screening and drug identification. A challenge in forensic/clinical toxicology is screening a vast number of analytes with a minimal sample pre-treatment and a speedy systematic step. TOF-HRMS, in combination with GC or LC is a better choice for screening purposes to identify designer steroids in urine. The HRMS methods incorporating hyphenation of SPE and LC-TOF has been developed to analyze many stimulants, narcotics, β-blockers, and β2-adrenergic agonists anti-estrogens, diuretics, and cannabinoids and metabolites in urine (Georgakopoulos et al., 2007; Kolmonen et al., 2007; Nielsen et al., 2010; Wójtowicz and Wietecha-Posłuszny, 2019).

Odoardi et al. (2017) have screened hair for illegal drugs and pharmaceutical drugs by LC-HRMS. The method was applied on 300 authentic hair samples from forensic cases (driving license renewals after driving under influence of drugs (DUID) occurrence, post-mortem toxicological analyzes, workplace drug testing, drug-facilitated crimes (DFC) (Odoardi et al., 2017). In the study, at least 30 mg, hair samples were collected by cutting in the posterior vertex zone by scissors, from maximum near to the scalp, and stored at room temperature in paper envelopes before the analysis. In a similar way, 10 drug free samples were also collected. The samples were washed 3 times, 2 mins each, vortexing with 1 mL of 0.1% solution of tween 80 (v/v in distilled water), rinsed two times with 0.5 mL of distilled water and once with 0.2 mL of acetone for 10 seconds, dried, and cut into small pieces with scissors. About 30 mg of each sample were added with 10 μL of deuterated internal standards (ISs) mixture (1 μg/mL) and left at 40°C overnight under sonication after addition of 300 μL of methanol. About 100 μL of the supernatant was then put in a micro vial and 10 μL was injected in LC-HRMS.

The samples registered the presence of various benzodiazepines, antidepressants, and antipsychotics (escitalopram, clotiapine, haloperidol, quetiapine, clomipramine, imipramine, mirtazapine, aripiprazole, carbamazepine, levomepromazine, trazodone) drugs of abuse and opioids (cocaine (COC), benzoylecgonine (BE), EME, cocaethylene, morphine, O-6-MAM, methadone, EDDP, MDMA, MDA, methamphetamine (MA), THC, oxycodone, synthetic cannabinoids, cathinone, ketamine (KET), ephedrine). Figure 8.4 shows the extracted ion chromatogram of an authentic hair sample showing the presence of clotiapine, aripiprazole, nordiazepam, and diazepam.

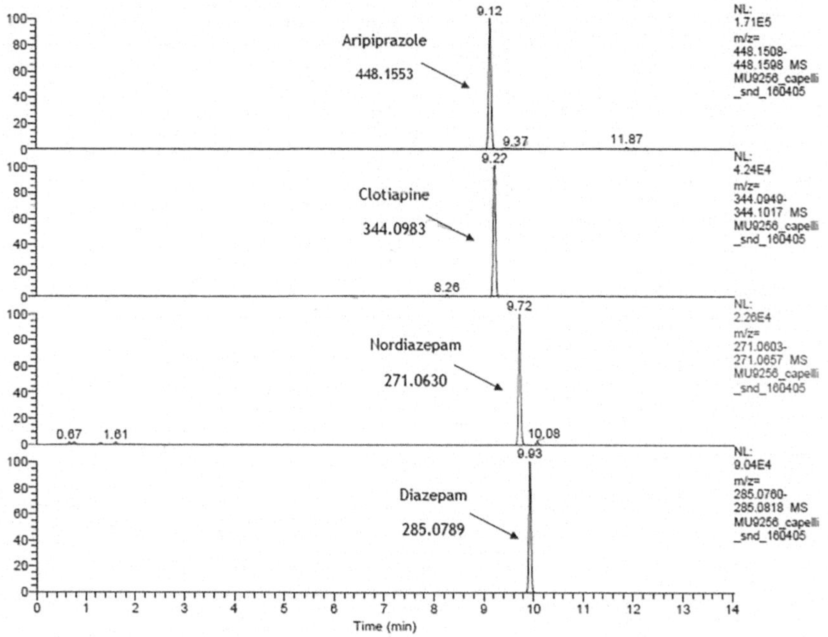

FIGURE 8.4 Extracted ion chromatogram of an authentic hair sample showing the presence of aripiprazole, clotiapine, nordiazepam, and diazepam.

Source: Reprinted with permission from Odoardi et al., 2017. © Elsevier.

Some target analytes were monitored from a single extraction and analysis, which include cannabinoids, amphetamines (AP), and AP-like substances, COC, KET, and analogs, both natural and synthetic cathinones, opiates, piperazines, ephedrines, and other new psychoactive substances not of these classes like pharmaceutical drugs such as anorectics, anti-epileptics, benzodiazepines, and analogs, stimulants, antidepressants, antipsychotics, and drugs for the erectile dysfunction. Among the methodologies applied for detecting analytes in complex matrices such as drugs in keratins, HRMS offers high selectivity for the target compound due to its exceptionally efficient mass resolution.

Richter et al. (2019) have analyzed antihypertensive drugs present in the oral fluid (OF) for monitoring drug adherence by using an LC-HRMS/MS method. A major problem faced in the hypertension treatment is the non-adherence to antihypertensive drug therapy. LC coupled to MS was shown to assess such non-adherence in blood and urine samples. A subjective method for the monitoring of adherence of 78 cardiovascular drugs in OF by using LC high resolution MS/MS (LC-HRMS/MS) was developed and validated (Richter et al., 2019).

Human OF, blood, and urine samples as well as the medication plans were collected from volunteers at Saarland University Hospital, Homburg, Germany, who were suffering from hypertension. Quantisal tubes were used for OF sample collection according to the manufacturer's instructions. Afterward, tubes were shaken for 2 h at room temperature, and cotton pads were removed. After centrifugation, the supernatants were collected and stored at –20°C. Matched ethylene diamine tetraacetate (EDTA) blood samples were centrifuged, and the separated plasma was stored along with the corresponding urine samples at –20°C until analysis. Blank OF was collected from voluntary people claiming to be drug-free. A 450 mL aliquot of the mixture was precipitated by 900 mL ice-cold acetonitrile and stored for 5 min at –20°C. The samples were centrifuged for 2 min and the supernatant was collected in a glass vial and was evaporated under nitrogen to dryness at 70°C. The residues were reconstituted in 75 mL methanol and again stored for 5 min at –20°C before being centrifuged for 2 min. From the supernatant, an aliquot was injected in the LC-HRMS/MS system. Samples of every patient were analyzed using the LC-HRMS/MS settings for the targeted adherence monitoring method. Figures 8.5 and 8.6 show the recovery and matrix effects of 78 cardiovascular substances in OF, detected by taking the advantage of full scan mode of LC-HRMS//MS. The developed subjective method was justified and used for the detection of antihypertensive medication in OF.

8.4 STUDIES OF PESTICIDES USING HRMS

People have been widely opened to several types of broad-spectrum substances due to rapidly evolving technology in current years. Pesticides have been a necessary part of agriculture to safeguard crops and livestock from pests and increase yield for many decades. However, its benefit for agriculture, pesticides could pose high risks to the environment, food safety, and human beings. There has been a promotion in the research of these pesticide agents' environmental fate, emigrating from treated fields to air, land, and water bodies, on account of the increased concern about the environmental and human impact of the pesticide applications. The necessity of pesticides in agricultural practices for developing countries is unavoidable; however, human health and environmental risks have emerged as a critical problem for all the countries, as reported by several studies. In the past 50 years, the pesticides usage contributed not only to the increase in agricultural production, but also increased environmental pollution and affects soil fertility (Özkara et al., 2016).

Deng et al. (2019) developed a new methodology for the analysis of perfluorinated compounds (PFCs) and fluorine containing pesticides (FCPs) present in environmental and drinking water by applying a supramolecular solvent microextraction followed by UPLC-Q-orbitrap HRMS analysis. Orbitrap HRMS is an attractive instrument for trace analysis of compounds in complex matrices. Therefore, the study aimed to develop and optimize a selective and sensitive analytical method, based on SUPRAS based microextraction and UPLC-Q-Orbitrap HRMS, to concomitantly determine four PFCs and five FCPs in water samples as shown in Figure 8.7.

Drinking water samples mainly include bottled water and barreled water, which were purchased from local markets. Environmental water samples include tap water acquired from laboratory Guangzhou, China), river water collected from the Zhujiang River (Guangzhou, China) and the wastewater obtained from some factories. The environmental water was filtered through 0.45 μm filter paper. The samples were stored at 4°C before analysis and analysis is carried out within 7 days of sample collection.

A rapid and effective UPLC-Q-Orbitrap HRMS method was applied for both the qualitative and quantitative examination of four PFC's and five FCP's in water samples. Figure 8.8 shows the extracted ion chromatograms of the nine compounds' mixed standards under the optimum conditions of UPLC-Q-Orbitrap MS.

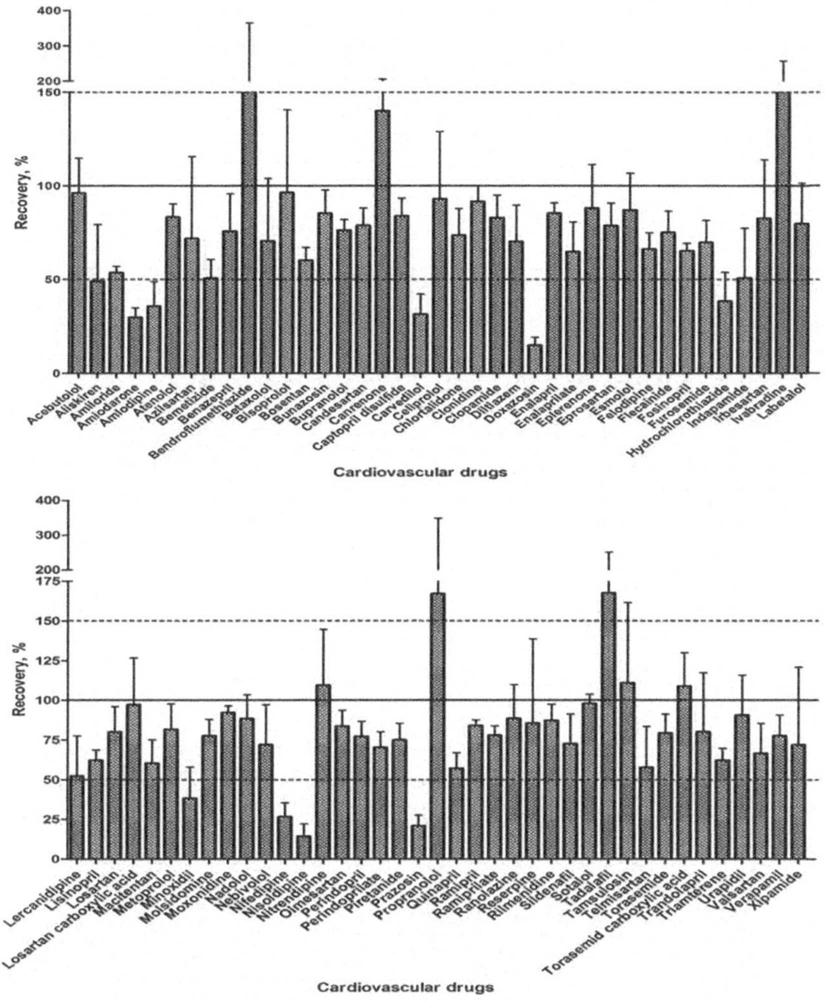

FIGURE 8.5 Recovery of the 78 cardiovascular substances in oral fluid, determined using full scan LC-HRMS/MS data.

Source: Reprinted with permission from Richter et al., 2019. © Elsevier.

Kiefer et al. (2019) have followed a distinct protocol of target and suspect screening for the detection of pesticide transformation products (TP) in groundwater with the aid of LC-HRMS. In order to examine the effect of pesticides on the quality of groundwater, about 31 groundwater samples, predominantly from intensive agriculture areas were subjected to the screening with a known target of 300 pesticides and suspect screening for more than 1,100 pesticide TP (Kiefer et al., 2019).

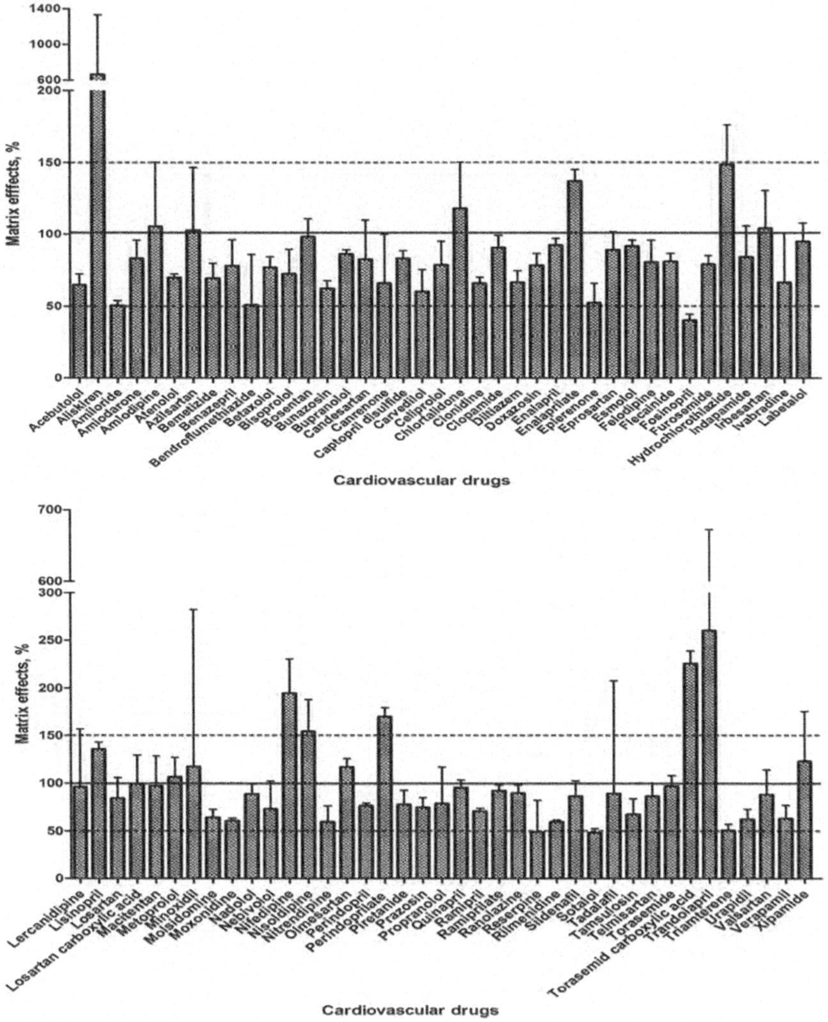

FIGURE 8.6 Matrix effects of the 78 cardiovascular substances in oral fluid, determined using full scan LC-HRMS/MS data.

Source: Reprinted with permission from Richter et al., 2019. © Elsevier.

As part of the routine sampling of Swiss National Groundwater Monitoring (NAQUA), groundwater samples were collected and were filled in glass bottles (which were previously annealed at 500°C; 500 mL and 1,000 mL bottles, SIMAX Kavalier, Czech Republic), stored for up to one week at <10°C in the dark and then kept frozen at –20°C until measurement. In order to ensure any possible contamination from the handling of samples, the blank samples were also prepared from ultrapure water, treated exactly the same as the samples and analyzed. The enrichment was achieved by concentration through evaporation under vacuum using the Syncore Analyst, in order to assure the protection of very polar compounds during enrichment.

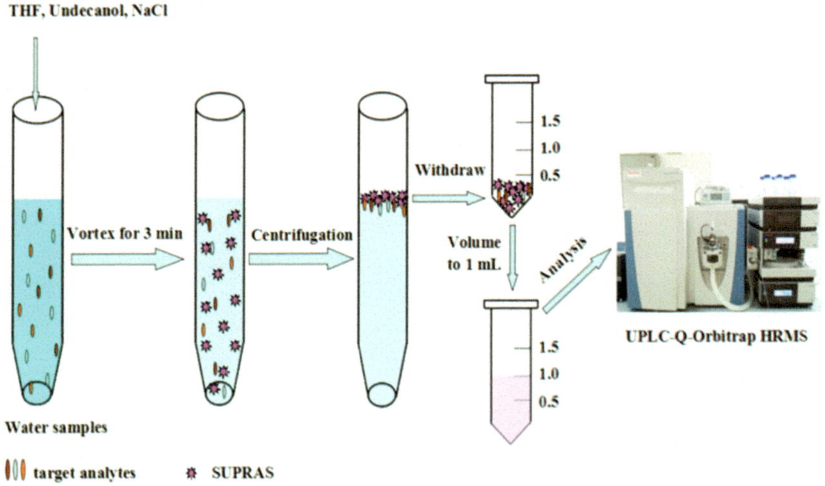

FIGURE 8.7 Schematic illustration of the SUPRAS based microextraction process.
Source: Reprinted with permission from Deng et al., 2019. © Elsevier.

Screening with a high-quality suspect list containing more than 1,000 experimentally observed pesticide TPs revealed the occurrence of several so far overseen pesticide TPs in groundwater and as shown in Figure 8.9. The suspect screening identified 27 pesticides and pesticide TPs (Levels 1–3). TPs of Chlorothalonil were found to present in groundwater in a wide manner and were detected in all the samples, including aquifers, which were supposed to have a low anthropogenic impact.

In Figure 8.9, land use in each groundwater monitoring site's catchment is given in the center. An upper left pie chart with the distribution of all quantitatively analyzed substances in compound classes is depicted. Suspects of pesticides and pesticide TPs were confirmed and quantified (Levels 1–3) or rejected. Agricultural micropollutants (MP): pesticides and suspects. Urban MPs: sweeteners, industrial chemicals, pharmaceuticals, and others. "X" marks groundwater monitoring sites close to wastewater impacted streams (<700 m, riverbank filtration). Monitoring sites are sequenced in the order of increasing agricultural land use, from top to bottom.

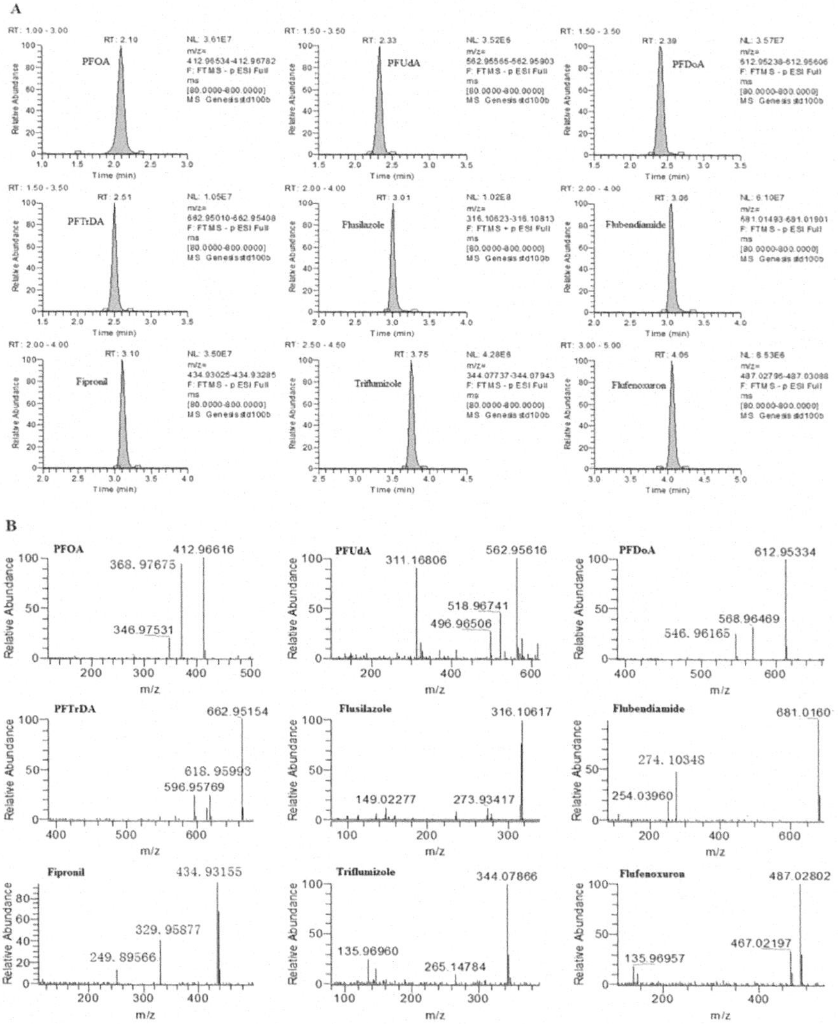

FIGURE 8.8 The extracted ion chromatograms (A); and MS2 chromatograms (B) of mixed standards of target compounds.

Source: Reprinted with permission from Deng et al., 2019. © Elsevier.

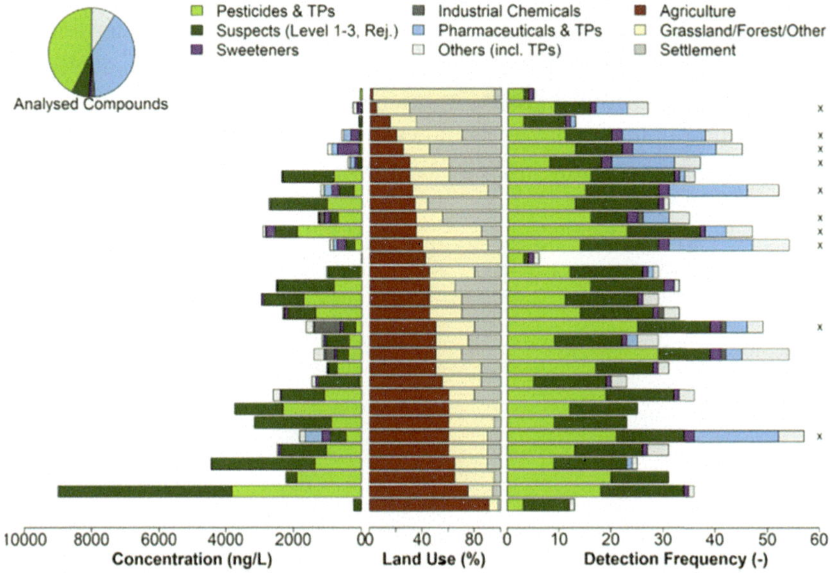

FIGURE 8.9 Total concentration (left) and detection frequency (right) per compound class and sample compared to land use.

Source: Reprinted with permission from Kiefer et al., 2019. © Elsevier.

8.5 TARGETED AND NON-TARGETED POLLUTANT

Organic pollutants are becoming an alarming challenge in the fields of environmental sciences, engineering, regulation, etc., for their vast and diverse applications. The utilization of HRMS, for non-target screening (NTS) s found to have enormous potential to characterize this world Hollender et al. (2017). The potential of hybrid quadrupole time of flight (QTOF), connected with ultra-high performance liquid chromatography (UHPLC) for the extensive screening of different types of samples, for organic contaminants were emphasized by Ramon Diaz et al. (2012).

Hollender et al. (2017) focused on the recent NTS application with HRMS coupled to chromatography. They showed that it was helpful to help solve real-world problems, opening up many opportunities to characterize processes and identify unknown pollutants. Figure 8.10 shows the workflow for NTS of typical environmental samples.

The most important river in central Europe is The Rhine, Basel. The Rhine station is responsible for daily monitoring river water quality (including detection and the long-term trend monitoring of accidental spills originating from industry, municipalities, or agriculture) to protect downstream populations. The routine screening analysis with LC-HRMS, GC-MS is the monitoring strategy as shown in Figure 8.11, so that downstream drinking water suppliers can be warned of spills in the same day itself and can take necessary actions, to avoid/stop the spills reach the drinking water extraction wells.

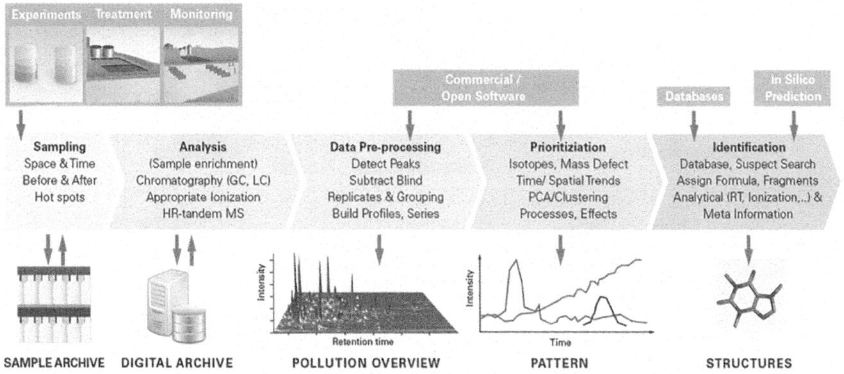

FIGURE 8.10 Workflow for a nontarget screening of environmental samples.

Source: Reprinted with permission from Hollender et al., 2017. © American Chemical Society.

Vanryckeghem et al. (2019) have reported a methodology, following Speedisk extraction and subsequent Q-orbitrap HRMS analysis for the screening of emerging organic MP as well as a multi-residue quantification. The methods were validated and were successfully used for tackling the challenge of quantifying the ultra-trace concentrations of organic MP present in seawater. This strategy was demonstrated successfully by applying to 18 samples collected from different locations in the harbor and open sea of the BPNS. Among the 89 target compounds, 63 and 17 compounds could be analyzed either in the positive or negative ionization mode, respectively. The remaining nine compounds (i.e., acyclovir, bezafibrate, chloridazon, diclofenac, diuron, efavirenz, imidacloprid, indomethacin, and naproxen) could be analyzed in both ionization modes leading to a total of 98 combinations of a compound, which were used to evaluate the optimization of the extraction protocols and validation parameters. Overall, 48 compounds could be detected on at least one sampling day and location, while 41 compounds remained undetected. Measured compounds include 24 pharmaceuticals, 5 PCPs and 19 pesticides with concentrations ranging from 0.1 to 156 ng/L for pharmaceuticals, 0.6 to 13 ng/L for PCPs, and 0.1 to 65 ng/L for pesticides. The 48 detected compounds included several personal care products, pharmaceuticals, pesticides, etc. The intimidating factor is that some of them were detected at concentrations ranging up to 156 ng/L for the first time in seawater.

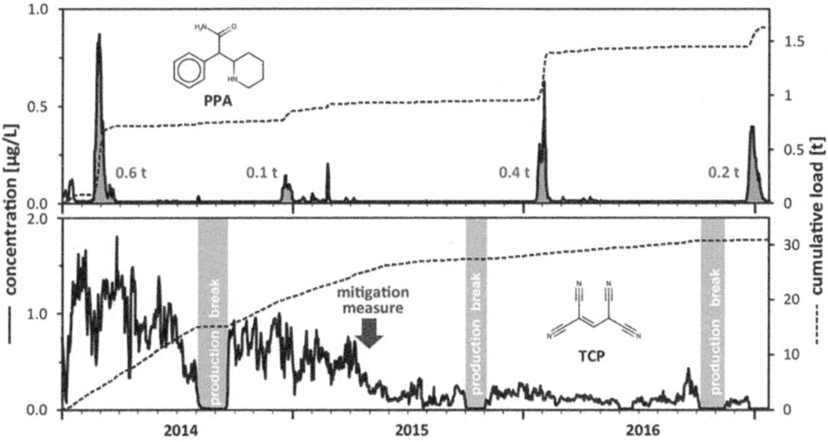

FIGURE 8.11 Results of daily LC-HRMS measurements between 2014 and 2016 at the River Rhine monitoring station close to Basel. (Top) Discharge events of the confirmed nontarget 2-phenyl-2-(2-piperidinyl)acetamide (PPA, $C_{13}H_{18}N_2O$, InChIKey LJLMNWPXAYKPGV-UHFFFAOYSA-N), each lasting several weeks, released over 1.6 tons of PPA into the Rhine between 2014 and 2016 from one production site. PPA is a precursor in the synthesis of methylphenidate (Ritalin). (Bottom) Continuous loads of a synthetic byproduct, confirmed as tetracarbonitrile-1-propene (TCP, $C_7H_2N_4$, InChIKey KPQNHJKRZKOLGW-HFFFAOYSA-N) from a single production site amounted to 31 tons. TCP emissions were reduced to zero during production breaks; process optimizations reduced emissions in 2015.

Source: Reprinted with permission from Hollender et al., 2017. © American Chemical Society.

Moreover, the newly developed HRMS method's untargeted screening potential was highlighted by identifying almost 1,300 compounds from a single sample and elucidating molecular formulae and structure to those components. This method furnishes a platform for the analysis of known targeted analytes and a simultaneous possibility of screening the unlimited number of unknown components, which comprises the marine environment. This was demonstrated by the detection of more than a thousand compounds in a single sample. The added value of untargeted screening was highlighted by providing information on molecular differences in fingerprints by selecting and assigning molecular formulae to the most differentiating components between sampling locations and sampling days.

8.6 CONCLUSION

HRMS is a vital entity in environmental MS, proceeds on the possibilities in the last decades. The methodologies incorporating HRMS are found to be the most efficient strategies for the wide-scope evaluation of the pollutants present in the environment. It seems to be the best approach to "ideal" universal screening, one of the most urgent demands in environmental analysis and other fields, such as food safety and toxicology. The evolution of wide-scope screening will surely be a key feature of HRMS in the next half-decade. Nonetheless, there are some compounds, especially those are present in their ionic form, that may require specific methods because they cannot be included in multi-residue methods (MRMs), owing to their specific characteristics. These scenarios were altered by the recent developments in the high-resolution (HR) mass analyzers, which have improved the analysis modes and results of the environmental samples. The search for the all-in-one instrument will continue in the future because the possibility of hyphenations and different analysis modes provide the option for all the required features in a single instrument setup, which fulfills the qualitative and quantitative analysis, with possibilities of structural elucidation of unknowns with high sensitivity selectivity, accurate mass, fragmentation, and linearity. As the HRMS methodologies offer improvement revolutions in many parts of the analysis, we will definitely witness corresponding changes in the applications as well. Undoubtedly, it will include environmental studies, food safety, toxicology, and others (Hernández et al., 2012).

KEYWORDS

- **antibiotics**
- **drugs**
- **gas chromatography**
- **liquid chromatography**
- **non-targeted pollutant**
- **pesticides**
- **pollutants**
- **time of flight**

REFERENCES

Cook – Botelho, J. C., Bachmann, L. M., & French, D., (2017). Steroid hormones. In: Nair, H., & Clarke, W., (eds.), *Mass Spectrometry for the Clinical Laboratory* (pp 205–230). Elsevier.

Deng, H., Wang, H., Liang, M., & Su, X., (2019). A novel approach based on supramolecular solvent microextraction and UPLC-Q-Orbitrap HRMS for simultaneous analysis of perfluorinated compounds and fluorine-containing pesticides in drinking and environmental water; *Microchem. J., 151*, 104250.

Díaz, R., Ibáñez, M., Sancho, J. V., & Hernández, F., (2012). Target and non-target screening strategies for organic contaminants, residues and illicit substances in food, environmental and human biological samples by UHPLC-QTOF-MS. *Anal. Methods, 4*, 196–209.

Georgakopoulos, C. G., Vonaparti, A., Stamou, M., Kiousi, P., Lyris, E., Angelis, Y. S., Tsoupras, G., et al., (2007). Preventive doping control analysis: Liquid and gas chromatography time-of-flight mass spectrometry for detection of designer steroids. *Rapid Commun. Mass Spectrom., 21*, 2439–2446.

Hernández, F., Sancho, J. V., Ibáñez, M., Abad, E., Portolés, T., & Mattioli, L., (2012). Current use of high-resolution mass spectrometry in the environmental sciences. *Anal. Bioanal Chem., 403,* 1251–1264.

Hollender, J., Schymanski, E. L., Singer, H. P., & Ferguson, P. L., (2017). Nontarget screening with high resolution mass spectrometry in the environment: Ready to go? *Environ. Sci. Technol., 51*, 11505–11512.

Hu, Q., Noll, R. J., Li, H., Makarov, A., Hardman, M., & Graham, C. R., (2005). The orbitrap: A new mass spectrometer. *J. Mass Spectrom., 40*, 430–443.

Ibáñez, M., Guerrero, C., Sancho, J. V., & Hernández, F., (2009). Screening of antibiotics in surface and wastewater samples by ultra-high-pressure liquid chromatography coupled to hybrid quadrupole time-of-flight mass spectrometry. *J. Chromatogr. A., 1216*, 2529–2539.

Kiefer, K., Müller, A., Singer, H., & Hollender, J., (2019). New relevant pesticide transformation products in groundwater detected using target and suspect screening for agricultural and urban micropollutants with LC-HRMS. *Water Res., 165*, 114972.

Kolmonen, M., Leinonen, A., Pelander, A., & Ojanpera, I., (2007). A general screening method for doping agents in human urine by solid-phase extraction and liquid chromatography/time-of-flight mass spectrometry. *Anal Chim Acta, 585*, 94–102.

Lebedev, A. T., Polyakova, O. V., Mazur, D. M., & Artaev, V. B., (2013). The benefits of high resolution mass spectrometry in environmental analysis. *Analyst, 138*, 6946–6953.

Lee, H. J., Ryu, H. D., Chung, E. G., Kim, K., & Lee, J. K., (2020). Characteristics of veterinary antibiotics in intensive livestock farming watersheds with different liquid manure application programs using UHPLC-q-orbitrap HRMS combined with on-line SPE. *Sci. Total Environ., 749*, 142375.

Nielsen, M. K. K., Johansen, S. S., Dalsgaard, P. W., & Linnet, K., (2010). Simultaneous screening and quantification of 52 common pharmaceuticals and drugs of abuse in hair using UPLC-TOF-MS. *Forensic Sci. Int., 196*, 85–92.

Odoardi, S., Valentini, V., De Giovanni, N., Pascali, V. L., & Strano-Rossi, S., (2017). High-throughput screening for drugs of abuse and pharmaceutical drugs in hair by liquid-chromatography-high resolution mass spectrometry (LC-HRMS). *Microchem. J., 133*, 302–310.

Özkara, A., Akyıl, D., & Konuk, M., (2016). Pesticides, environmental pollution, and health. In: Larramendy, M. L., (ed.), *Environmental Health Risk-Hazardous Factors to Living Species*. IntechOpen: UK.

Richter, L. H., Jacobs, C. M., Mahfoud, F., Kindermann, I., Böhm, M., & Meyer, M. R., (2019). Development and application of a LC-HRMS/MS method for analyzing antihypertensive drugs in oral fluid for monitoring drug adherence. *Anal. Chim. Acta, 1070*, 69–79.

Vanryckeghem, F., Huysman, S., Van, L. H., Vanhaecke, L., & Demeestere, K., (2019). Multi-residue quantification and screening of emerging organic micropollutants in the Belgian part of the North Sea by use of speedisk extraction and Q-orbitrap HRMS. *Mar. Pollut. Bull., 142*, 350–360.

Wójtowicz, A., & Wietecha-Posłuszny, R., (2019). DESI-MS analysis of human fluids and tissues for forensic applications. *Applied Physics A, 125*, 312.

Wu, A. H. B., Gerona, R., Armenian, P., French, D., Petrie, M., & Lynch, K. L., (2012). Role of liquid chromatography-high-resolution mass spectrometry (LC-HR/MS) in clinical toxicology. *Clin. Toxicol., 50*, 733–742.

CHAPTER 9

Single Cell Analysis Methods Based on Mass Spectrometry

RESHMA V. R. NAIR

R&D Center, Aurea Biolabs (P) Ltd., Kolenchery, Cochin – 682311, Ernakulam, Kerala, India, E-mail: reshmavr1992@gmail.com

ABSTRACT

Cells are the basic structural and functional units in biological systems that play important roles in living organisms. A single cell is mainly composed of a myriad of biochemical compounds with enigmatic regional roles. The single-cell analysis provides the precise mechanism of cellular and molecular behavior in the cell and provides unique analytical capabilities for biological and biomedical research, including determination of the biochemical heterogeneity of cellular populations and intracellular localization of pharmaceuticals, study of genomics, epigenome, transcriptomes, proteomics, and metabolomics at the single-cell level. Various analytic techniques (flow cytometry, fluorescence microscopy, Raman microscopy, and mass spectrometry (MS)) are used for single-cell analysis and each technique has its own advantages and disadvantages. Among them, MS has been widely used in single-cell analysis in recent years due to its high sensitivity, high specificity, capability for structure identification, quantify chemicals, low sample consumption with high detection potential of metabolites present in a single cell. ESI-MS, LDI-MS, SIMS, and ICP-MS are the MS techniques most often used in micro-bioanalytical investigations of single cells. Nowadays, researchers continuing advancements in

High-Resolution Mass Spectrometry and Its Diverse Applications: Cutting-Edge Techniques and Instrumentation. Sreeraj Gopi, PhD, Sabu Thomas, PhD, Augustine Amalraj, PhD, & Shintu Jude, MSc (Editors)

MS, which provide higher spatial resolution and sensitivity in single-cell analysis. This chapter tries to look at the recent advances in single-cell analysis and high resolution MS imaging of single cells.

9.1 INTRODUCTION

A single cell is the fundamental unit of biological organisms. Humans are complex organisms made up of almost trillions of cells, each cell work in harmony to carry out all the basic functions necessary for humans to survive. Stem cells, liver cells, skin cells, muscle cells, blood cells, neurons, etc., are the important cells found in our body. The ability to study, analyze, and manipulate each individual cell has been the focus of a lot of contemporary research and which is an emerging field in cell biology in the recent years. So single-cell analysis is a critical tool for understanding cellular heterogeneity, study of genomics, epigenome, transcriptomes, proteomics, and metabolomics at the single-cell level. It reduces the molecular noise obtained from a large heterogeneous cell population and discovering the unique characteristics of individual cells. Chemical characterization of these single cells was very challenging target to researchers over the years and they have developed alternative methods to visualize or image tissues or cells; these include observing the compound-specific vibration or resonance with spectroscopy techniques such as UV, IR, anti-Stokes Raman spectroscopy, Nuclear magnetic resonance spectroscopy (NMR) or directly measuring a compound's mass with Mass spectrometry (MS) (Svatos, 2011).

In the 20th century, to analyze a single cell metabolite directly, scientists have introduced one of the most successful standard operational procedure MS with tandem MS (MS/MS) for imaging and profiling which provide distinctive analytical capabilities for biological and biomedical research (Lanni et al., 2012). The inherent sensitivity, high specificity, and capability for structure identification of mass spectrometry imaging (MSI) make it a desirable detector for the analysis of different metabolites in a single cell.

MSI of single-cell mainly performed through secondary ion mass spectrometry (SIMS), Matrix-assisted laser desorption ionization (MALDI), electrospray ionization (ESI), and inductively coupled plasma mass spectrometry (ICP-MS) (Figure 9.1). Due to the immense importance of single-cell analysis, our research community has devised various methods for the chemical characterization and identification of single-cell by high-resolution mass spectrometry (HRMS). Here we discussed various techniques and research based on single-cell analysis by HRMS.

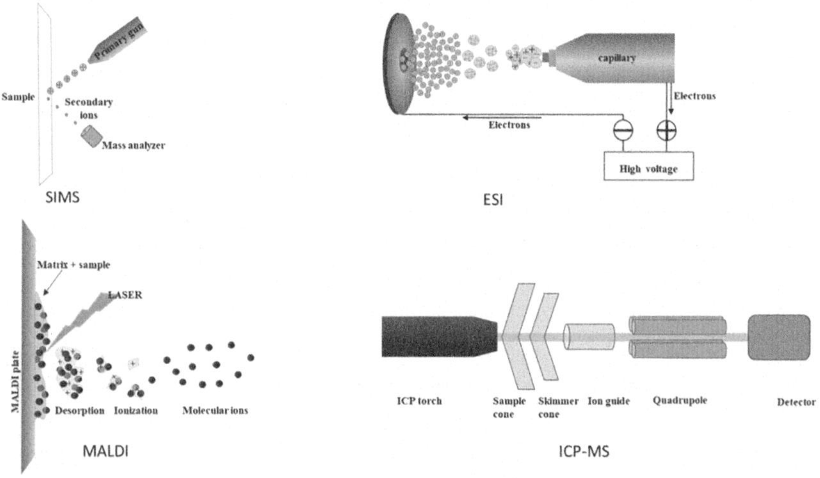

FIGURE 9.1 Schematic representation of different mass spectroscopic techniques (SIMS, ESI, MALDI, ICP-MS).

9.2 SECONDARY ION MASS SPECTROMETRY (SIMS)

SIMS was developed in the 1960s. It was the first MS imaging technique applied to chemical imaging and has been used to image a broad range of biological specimens, including single cells. SIMS is based on the principle that an energetic primary ion beam bombards the surface and generates secondary ions in order to induce desorption of intact molecules, fragments, and atoms from the first few nanometers of the sample surface. Secondary ions of a single polarity are extracted and subsequently transferred into a mass analyzer, usually a magnetic and/or electrostatic sector or a time-of-flight (TOF) analyzer. A lateral resolution better than 100 nm is possible with specialized SIMS instruments. SIMS has the ability to observe the distributions and measure the amount of each species within complex biological samples and within a specified region of the sample.

It is the only currently available method capable of high resolution (HR) three-dimensional (3D) imaging and has important application value in the field of life sciences, especially single-cell imaging analysis.

It is a relatively hard ionization technique compared with other ionization modes such as electron spray ionization (ESI) or MALDI. Thus, the characteristics which provide true analysis of the sample, although many challenges must be resolved with the introduction of SIMS. One of the major difficulties for the target molecules analysis is high-energy ion beam generates many fragment ions which restricted SIMS in the analysis of single-cell biological macromolecules. So nowadays researchers mainly focused on the development of softer SIMS ion source and the efficient ionization of larger molecular mass single cells. The emergence of gas cluster ion beams (GCIB) techniques has a softer SIMS ion source which generate lesser ion fragments for the ionization of higher molecular mass single biomolecules (Ju-Duo et al., 2020). In the analysis of SIMS, two fundamentally different approaches have been developed – TOF-SIMS and dynamic SIMS using a magnetic sector mass spectrometer (SECTOR).

For static TOF-SIMS, the primary ion beam is bombarded and the obtained secondary ions are detected by a TOF mass spectrometer. It provides a full mass spectrum to be monitored during the analysis. The goal of TOF-SIMS is the ejection and detection of molecular species. The only less material is needed for the ionization and detection of the sample due to its static effect which is suitable for the study of intact molecular ions and larger biomolecular fragments such as peptides (Komatsu et al., 2008). With microprobes focusable to a submicron diameter of range above 50 nm with dynamic SIMS due to the surface availability of the analyte, TOF-SIMS provides relatively more chemical information at lower diameter at subcellular lateral resolutions. Dynamic SIMS is performed with a continuous primary ion beam is bombarded and a preselected set of ions is detected with a magnetic SECTOR (Boxer et al., 2009). Dynamic SIMS otherwise called magnetic sector SIMS, in which reactive primary ions enhance the secondary ion yields. Due to the continuous bombardment of ions, molecular bonds are broken, and only small molecular and monoatomic ions are produced and detected by this instrument.

Tian et al. (2017) reported the development of 3D imaging cluster TOF-SIMS as a label-free approach to chemical imaging the small molecules in aggregated and single *Escherichia coli* cells, with ~300 nm spatial resolution and a depth resolution of 200 nm as well as high chemical sensitivity.

They used C_{60}^+ ion beam to perform subcellular imaging analysis of the distribution of two antibiotics viz. ampicillin (AMP) and tetracycline (TET), in *E. coli*. The results were consistent with the result of fluorescence confocal microscope imaging and immunoblot analysis (Tian et al., 2017).

Pareek et al. (2020) used 70 keV $(CO_2)_n^+$ (n > 10,000) ion beam to perform *in situ* 3D submicrometer chemical imaging of single cells by gas cluster ion beam secondary ion mass spectrometry (GCIB-SIMS) to directly visualize de novo purine biosynthesis by a multienzyme complex, the purinosome in Hela cell. They found that the purinosome responsible for purine synthesis in the cell comprise nine enzymes that act synergistically. The purinosome regulated the synthesis of purine nucleotides by controlling the synthesis of purines and regulated the adenosine monophosphate/guanosine monophosphate ratio in the cell (Pareek et al., 2020).

Sheng et al. (2017) combined SIMS with stable isotope (^{15}N) labeling technique for visualization of newly synthesized proteins during the cell growth. They simultaneously image phosphatidylcholine and cholesterol during the cell division, since TOF-SIMS is especially suitable for this kind of molecules. They cultured Hela cell using a medium containing ^{15}N-arginine and ^{15}N-lysine and after incubation, the ^{15}N-labeled amino acids and phospholipids in the cell were imaged and analyzed by this technique. They proposed that phosphatidylcholine was widely distributed in the division area while the newly synthesized proteins was exist less in this region (Sheng et al., 2017).

Huang et al. (2016) proposed high-throughput single-cell patterning method, which was firstly developed for the preparation of single-cell arrays. This method combined polydimethylsiloxane (PDMS) material micro-processing technique and centrifugation-assisted cell trapping techniques for the preparation of surface single-cell microarrays. The occupancy rate of cells in the microarray site could reach more than 90%, of which the single-cell resolution rate was more than 97%. Then they combined this single-cell array preparation method with TOF-SIMS to achieve high-throughput preparation and analysis of single-cell samples, and successfully investigate it to the study of drug-induced early phenotypic changes of Hela cells in apoptosis at both population and single-cell levels (Huang et al., 2016).

Hua et al. (2018) reported delayed extraction mode of TOF-SIMS to analyze the silver nanoparticles induced lipids changes on the surface of single macrophage cells. TOF-SIMS provided high mass resolution of mass spectra and high spatial resolution of single-cell chemical imaging. In this work they used TOF-SIMS to investigate silver nanoparticles induced lipids changes on the surface of single macrophage cells. Studies showed that there is a significant distinction observed between cells treated with different concentrations of silver nanoparticles had differences in the distribution of surface lipids, mainly including cholesterol, monoglycerides, and diglycerides. These results demonstrated the feasibility of TOF-SIMS for characterizing the changes of the lipids on the single-cell surface and providing a better interaction between nanoparticles and cells (Hua et al., 2018).

Liu et al. (2017) combined SIMS-TOF with confocal laser scanning microscopy imaging technique and successfully applied to visualize the subcellular distribution of an organoruthenium anticancer complex. They studied to achieve imaging analysis of an organoruthenium anticancer complex drug in a single cancer cell. The imaging results showed the organoruthenium complexes mainly concentrated in the cell membrane and nucleus and the results were more accurate and reliable. This method had great application in the field of organometallic drugs and pharmacological research (Liu et al., 2017).

Passarelli et al. (2017) reported the development of a 3D OrbiSIMS instrument for label biomedical imaging. They combined the high spatial resolution 3D SIMS (under 200 nm for inorganics and under 2 μm for biomolecules) with the high-quality resolution of Orbitrap (> 240,000 at m/z 200) to realize visual 3D imaging analysis of endogenous and exogenous cellular metabolites at the subcellular level. They used the instrument to image and analyze metabolites such as gamma-aminobutyric acid (GABA), dopamine, and serotonin in the neurons of mouse hippocampus. The techniques provide critical insight on fundamental biological processes (Passarelli et al., 2017).

Monroe et al. (2005) reported the distribution of major constituents the cellular membrane of isolated neurons. The results show a significant increase in the vitamin E signal at the soma-neurite junction compared to a whole-cell and which provided an idea about the new role of vitamin E in the neuronal function (Monroe et al., 2005).

Court et al. (2017) reported another application of ION TOF-SIMS 5–100 (with a 25 keV Bi^{3+} primary ion beam) was performed to describe molecular changes of how various polarizations are modified by low-oxygen exposure of macrophage cells. They introduced some species, specificity of a particular polarization, advocating for the use of macrophages as reporter cells of oxygen tension in tissues. These results open a new way of TOF-SIMS to explore hypoxia in human tissues (Court et al., 2017).

Ostrowski et al. (2004) demonstrate a decreased abundance of phosphatidylcholine and an increase in an aminoethylphosphonolipid detected at the plasma membrane sites between fusing Tetrahymena. From the mass spectral data reports the fusion region contains elevated amounts of 2-aminoethylphosphonolipid, a high-curvature lipid (Ostrowski et al., 2004).

9.3 MATRIX-ASSISTED LASER DESORPTION-IONIZATION (MALDI)

Matrix-assisted laser desorption-ionization (MALDI) MS is a rapid and soft ionization analytical approach that is well suited for obtaining molecular weights of peptides and proteins from complex samples which provides the exciting possibility of studying the pattern of neuropeptide gene expression at and below the level of a single cell. Due to its high sensitivity, proteins, and peptides usually ionize as intact, singly protonated ions so that the accurate molecular weights of the peptides are easily acquired for the characterization of neuropeptides. The major goal of MALDI MS is to vaporize and ionize the analytes without fragmenting them (i.e., without introducing excess energy), by incorporating the analyte into an excess of a matrix. The matrix absorbs energy from a pulsed laser, vaporized, and ionized. Following gas-phase ion generation, a mass analyzer and detector is needed to acquire the mass spectra (Li and Sweedler, 2000).

With the innovation of MALDI-based sequencing techniques, single-cell MALDI approaches have grown from simple molecular weight profiling of previously identified peptides to well-designed experiments that enable the discovery and structural elucidation of novel peptides. Compared with other conventional approaches for cellular peptide characterization, attractive features of MALDI-based single-cell profiling are compatibility with crude mixtures, efficient use of the limited amounts of analytes, minimal sample purification, label-free analysis, chemical specificity, simple instrumentation, high sensitivity (a lateral resolution of ~10 μm is currently attainable with commercial instruments), covering a wide range of molecules that can be visualized simultaneously and spatial distribution of metabolites (Boggio et al., 2011; Li and Sweedler, 2000; Ščupáková et al., 2020). Development of higher-efficiency matrix materials, the improvement of the ionization efficiency and detection sensitivity of the molecules to be tested are the mostly targeted field in MALDI research.

Kompauer et al. (2017) reported an atmospheric pressure (AP) matrix-assisted laser desorption/ionization (MALDI) MSI setup with a lateral resolution of 1.4 mm with a MALDI imaging analysis system based on a new ultraviolet laser focusing system. They achieved a mass resolution greater than 100,000, and accuracy below 62 ppm. The studies reported that the experimental setup can be used to detect metabolites, lipids, and small peptides, as well as to perform tandem MS experiments with 1.5-mm sampling areas. These analyses were performed through the imaging analysis of metabolites in different parts of cilia and oral groove of *Paramecium caudatum* at the subcellular level (Kompauer et al., 2017).

To recover the limitations of MALDI-TOF MS and TOF-SIMS Wang et al. (2018) introduced a new method, vacuum ultraviolet laser desorption/ionization (VUVDI) for improving the high spatial resolution, achieving higher ion yields and less fragmentation of single-cell imaging. They designed an optical path system which was used to focus the ultraviolet laser through an aspheric MgF_2 lens. The diameter of the single spot ablated on the sample surface was controlled at about 700 nm. From the results VUVDI-MSI with submicron spatial resolution has significant potential, and it became a powerful MSI tool for biological samples, especially for single cells in the future (Wang et al., 2018).

Haag et al. (1998) reports the use of MALDI/TOF-MS as a technique for the rapid identification and speciation of Haemophilus present in both pure and mixed cultures owing to its ability to detect fingerprint proteins. MALDI/TOF mass spectra can be acquired at 10 min, which allow the characterization of the species within 24 rather than the 48 h or more needed for traditional bacteriological methods and analysis can perform immediately after growth occurs (Haag et al., 1998).

van Veelen et al. (1993) reported the first peptide profiles of single neurons in *Lymnaea stagnalis* and were directly characterized by MALDI mass spectrometry. They performed mass analysis after minor pretreatment without any separation steps and compared the study with the results of conventional peptide chemical methods. These studies introduced known peptides and several new peptides and the method and results provides a new way to the study of peptide compositions at the single-cell level (van Veelen et al., 1993).

For informative peptide profiling on complex samples such as organs or cells and identification of specific peptides in biological extracts or cells, Redeker et al. (1998) combined MALDI-TOFMS and immunological tools (ELISA) and immunocytochemistry. They analyzed the cellular expression of the two precursors of the hyperglycemic hormone (cHH) in neurosecretory cells (30-μm diameter) from the crayfish *Orconectes limosus*. The analyzed samples were recovered from the MALDI plate for further immunological analysis by ELISA. The combined characterizations were scaled up to a single crayfish neurosecretory cell level (Redeker et al., 1998).

9.4 ELECTRON-SPRAY IONIZATION (ESI)

SIMS and MALDI are the emerging methods in single-cell analysis, but these methods are generally operated under vacuum conditions and are not suitable for live cell analysis. So there is an introduction of different ESI techniques has been developed to accomplish the single-cell analysis. The ESI-based method can detect sample cells within their natural environment and provide information on a less perturbed cell state (Stopka et al., 2018). Desorption electrospray ionization (DESI) is spray based ionization method and generates ion in native conditions of temperature and pressure. DESI is an improved version of the ESI, which is a soft ionization technique that converts ions into gaseous phase at AP. ESI was developed for the analysis of biological molecules and other purpose, including drug analysis, painting analysis and food analysis. DESI provides a spatial resolution ranging from 300 μm to less than 50 μm, and the best of DESI system achieve a resolution less than 10 μm (Nisar and Zhao, 2020). In some times due to the matrix effect reduction in the accuracy and precision of the results of single-cell analysis noted so as to overcome these difficulties nano-ESI-MS, probe ESI (PESI-MS), capillary ESI (CESI-MS) and laser ablation ESI (LAESI-MS) were introduced (Patel et al., 2020).

With the increasing demand for detecting single cell *in situ* and *in vivo*, laser ablation electrospray ionization mass spectrometry (LAESI-MS) was invented. LAESI-MS can sample individual cells in their native environment, and cell-by-cell molecular imaging of metabolites in plant epidermal tissue was demonstrated, where individual cells were manually selected to serve as natural pixels for tissue imaging. Imaging was carried out by manually moving the cells within the sample, however, it is not feasible for the analysis of numerous single cells or selected cell types in extended tissue areas (Li et al., 2015). LAESI-MS is well suitable for the detection of the water-containing analyte and has the potential for two-dimensional molecular imaging and depth profiling, and can be explored for the detection of biochemical changes in live tissues (Patel et al., 2020).

Nano-ESI is a soft ionization method with high ionization efficiency, which is based on nanotip sampling. The sample analyte is placed in a nanotip and a high voltage (1–2 kV) is applied to the sample solution through the electrode. Due to the effect of high voltage, the sample is ejected as an electron spay which is observed through a microscope and nanotip is controlled by a three-dimensional mobile manipulator to directly sample from the target area of the living cell. After sampling, direct ionization and detection can be performed through Nano ESI. Compared with other ESI ionization techniques, it has the advantages of direct, spontaneous live sampling, superior ionization efficiency and sensitivity and no special conditions (matrix and vacuum) are required during the test (Zhang et al., 2016).

Probe-ESIMS (PESI) concept was proposed by Hiaroka and for this technique an extra sample preparation step is required before sample injection. PESI provides a method to analyze the single cell by enriching and isolating metabolites/subcellular with specific biological significance or function from cells (Hiraoka et al., 2007). Different types of metabolites like fructan, lipid, and flavone derivatives have been analyzed from a single cell. PESI has its distinctive merits in single-cell analysis, such as direct sampling in live cells, enrichment of metabolites which improved the sensitivity, subcellular detection could be achieved with miniaturized probes, reduced disturbance to the original distributions of metabolites during sampling and simplified device and reusable probes. Significant enhancement (~30-fold) in the sensitivity was observed in this method and it is also one of the most widely used and effective methods for single-cell analysis (Gong et al., 2014).

While PESI based single-cell analysis is more applicable to the plant cells, we needed more sensitive methods for analysis of animal cells due to their smaller volume and the lower content of intracellular substances.

One of the most economical and convenient methods to improve the sensitivity and accuracy of the method are the preparation of functional probes to enrich trace components from a single cell and improving the ionization and transmission efficiency of trace components. So far, there are few studies have done on these two aspects, which projects the need for more research and studies on the PESI method. Another improved electrospray ionization method is hydrogen flame desorption ionization mass spectrometry (HFDI-MS), which has been successfully applied to the quantitative analysis of the single-cell level (Zhao et al., 2019).

Nakashima et al. (2016) used another approach for the MS analysis of single stalk and glandular cells, two adjacent cell types in intact trichomes (small often single-cell hair-like growths from plant cells) of tomato plants (*Solanum lycopersicum*) using the pressure probe electrospray ionization-mass spectrometry with internal electrode capillary (IEC-PPESI-MS). Studies of this group revealed high spatial-resolution cell sampling, precise post sampling manipulation, and high detection sensitivity of this method. They used this method for less than a picoliter cell sap from a single stalk cell, which have sufficiently yielded a number of peaks of amino acids, organic acids, carbohydrates, and flavonoids. This study shows that me metabolites were found only in specific cell types or particular trichome types (Nakashima et al., 2016).

Yin et al. (2018) demonstrated quantitative mass spectrometry analysis of biomolecules in picoliter volumes extracted from live cells. They achieved the extraction using electroosmotic drag and detection of several metabolites, including sugars and flavonoids from the cytoplasm of individual Allium cepa cells using two electrodes and a finely pulled nanopipette (with tip diameter of <1 μm) which was coupled to a nano ESI. They studied that the ratio of the signal of glucose normalized to the signal of glucose-d2 which increased linearly with glucose concentration (Yin et al., 2018).

Hu et al. (2016) introduced and applied a synchronized polarization induced electrospray ionization (SPI-ESI) method for dual-polarity MS analysis of single cells like PC-12 cells and *Alium cepa* cells. In SPI-ESI method, a periodic alternating current square wave voltage (AC-SWV) is applied to induce the bipolar spray and both positive and negative-ion mass spectra are obtained from one measurement through the bipolar spray process by synchronizing mass analyzer. Furthermore, an ultra-low spray flow rate was achieved (picoelectrospray ionization, pESI, flow rate

<1,000 pL/min) in SPI-ESI without loss of its sensitivity. They reported and profiled 86 components from single *A. cepa* cells and 94 components from single PC-12 cells, and it was the first study reports for the comprehensive analysis of single-cell samples by combining both positive-ion and negative-ion mass spectra (Hu et al., 2016).

Lombard-Banek et al. (2016) developed an instrument design (CE-μESI-HRMS) which combines capillary electrophoresis (CE), electrospray ionization (ESI), and a tribrid ultrahigh-resolution mass spectrometer (HRMS) to enable untargeted (discovery) proteomics with about 25amol lower limit of detection in individual embryonic cells. CE-μESI-HRMS enabled the identification of 500–800 nonredundant protein groups by measuring the total protein content in single blastomeres (a type of cell produced after cell division of the zygote after fertilization) that were isolated from the 16-cell frog (*Xenopus laevis*) embryo. The isolation is carried out using microdissection, followed by digestion and a CE-microflow electrospray ion source (μESI) leading to a high-resolution tandem mass spectrometer (Lombard-Banek et al., 2016).

Xu et al. (2019) established an electrosyringe-assisted ESI-MS method to achieve intracellular sampling from axons or dendrites in living neurons. They developed and inserted a capillary tip (~130 nm) into one axon or dendrite to extract cytosol through electro-osmotic flow. The mass spectrometry (MS) techniques show that the amounts of pyroglutamic acid and glutamic acid were higher in axons compared to the dendrites and cell body due to the accumulation of neurotransmitters in the axon for information delivery (Xu et al., 2019).

9.5 INDUCTIVELY COUPLED PLASMA MASS SPECTROMETRY (ICP-MS)

Metals and nonmetals (Fe, Mn, Zn, Cu, Ni, P, S, B, and Cl) like essential and important components of living systems which are present in the form of metalloproteins, metalloenzymes, and nucleic acids, and play very significant roles in cellular activities. ICP-MS includes attractive features for single-cell analysis, such as high sensitivity (ng L^{-1} range), wide linear dynamic detection range and specificity for the accurate detection and quantification of metals, metalloids, and heteroelements (including non-metals, semi-metals, and halogens). It does not depend upon the sample matrix or ion signal suppression effects, even in complex and heterogeneous biological specimens. By directing laser-ablated sample material into the plasma in a line-by-line fashion before the measurement with a carrier gas such as argon. ICP-MS have a high temperature process provides the efficient vaporization, dissociation or atomization, excitation, and final ionization of the detectable atomic constituents, which are released and analyzed with mass spectrum. Moreover, the detection of metabolites can be performed via metal-tagging followed by ICP-MS.

ICP-MS is classified into classical, time-resolved, and laser ablation (LA). In the-classical technique, after lysis, extraction, and digestion, the mass of the cell is taken for cellular analysis. So it is very complicated to get any significant information related to cell-to-cell variation between each cell through this technique. This limitation can be overcome with the introduction of time-resolved ICP-MS for single-cell analysis. In this technique, a cell suspension is passed to a plasma torch through a nebulizer or dispenser which followed the generation of the ion cloud. The obtained ion cloud can be analyzed as an individual signal with a high time resolution. In LA-ICP-MS technique, the samples are ablated through the laser light in a line-by-line fashion before the measurement which provides the high spatial resolution, sensitivity, and flexibility (Patel et al., 2020).

LA is often coupled with ICP-MS for two modes of single-cell analysis either for profiling or imaging of cell surfaces of a single cell. To detect the complex cellular systems and low-abundant cellular molecules at single-cell resolution, mass cytometry involving a fusion of flow cytometry with ICP-MS was developed (Oomen et al., 2018). This technique involves antibodies that are conjugated with isotopically pure elements, and are used to label cellular molecules (mostly proteins). Currently, mass cytometry is a useful tool which allows simultaneous measurement of over 40 cellular parameters at single-cell resolution with the throughput required to survey millions of cells from an individual sample (Spitzer and Nolan, 2016).

Managh and coworkers reported the first application of LA-ICPMS to perform single-cell detection of T cell populations relevant to cellular immunotherapy. They labeled purified human CD_4^+ T cells with commercially available Gd-based magnetic resonance imaging (MRI) contrast agents, Omniscan, and Dotarem, which enabled passive loading of up to 108 Gd atoms per cell. They invented that the gadolinium-labeled human cells retained measurable for up to 10 days both *in vitro* and *in vivo* in an immunodeficient mouse model (Managh et al., 2013).

9.6 CONCLUSION

Single-cell analysis is the emerging frontier in the field of omics, and which has the potential to transform systems biology through new discoveries derived from cellular heterogeneity. MS is an important analytical technique that has found pervasive use in detection and identification of a single cell. Continuing advancements and research in MS provide higher spatial resolution and sensitivity in single-cell analysis. In this chapter, we concisely explained the various MS techniques with their merits and demerits that are widely explored for single-cell analysis at cellular and subcellular levels. From the last decades, the research on single-cell MS has made very good progress in scientific and biomedical fields. Therefore, advanced MS methods and techniques are still needed for high-throughput profiling and imaging of metabolites using single cells.

KEYWORDS

- **electrospray ionization**
- **inductively coupled plasma mass spectrometry**
- **mass spectrometry imaging**
- **matrix-assisted laser desorption ionization**
- **secondary ion mass spectrometry**
- **single-cell analysis**

REFERENCES

Boggio, K. J., Obasuyi, E., Sugino, K., Nelson, S. B., Agar, N. Y., & Agar, J. N., (2011). Recent advances in single-cell MALDI mass spectrometry imaging and potential clinical impact. *Expert Rev. Proteomics, 8*, 591–604.

Boxer, S. G., Kraft, M. L., & Weber, P. K., (2009). Advances in imaging secondary ion mass spectrometry for biological samples. *Annu. Rev. Biophys., 38*, 53–74.

Court, M., Barnes, J. P., & Millet, A., (2017). Identifying exposition to low oxygen environment in human macrophages using secondary ion mass spectrometry and multivariate analysis. *Rapid Commun. Mass Spectrom., 31*, 1623–1632.

Gong, X., Zhao, Y., Cai, S., Fu, S., Yang, C., Zhang, S., & Zhang, X., (2014). Single-cell analysis with probe ESI-mass spectrometry: Detection of metabolites at cellular and subcellular levels. *Anal. Chem., 86*, 3809–3816.

Haag, A. M., Taylor, S. N., Johnston, K. H., & Cole, R. B., (1998). Rapid identification and speciation of *Haemophilus* bacteria by matrix-assisted laser desorption/ionization time-of-flight mass spectrometry. *J. Mass Spectrom., 33*, 750–756.

Hiraoka, K., Nishidate, K., Mori, K., Asakawa, D., & Suzuki, S., (2007). Development of probe electrospray using a solid needle. *Rapid Commun. Mass Spectrom., 21*, 3139–3144.

Hu, J., Jiang, X. X., Wang, J., Guan, Q. Y., Zhang, P. K., Xu, J. J., & Chen, H. Y., (2016). Synchronized polarization induced electrospray: Comprehensively profiling biomolecules in single cells by combining both positive-ion and negative-ion mass spectra. *Anal. Chem., 88*, 7245–7251.

Hua, X., Li, H. W., & Long, Y. T., (2018). Investigation of silver nanoparticle induced lipids changes on a single cell surface by time-of-flight secondary ion mass spectrometry. *Anal. Chem., 90*, 1072–1076.

Huang, L., Yin, C., Lu-Tao, W., Mark, L., Xiaoxing, X., Zhiyong, F., & Hongkai, W., (2016). Fast single-cell patterning for study of drug-induced phenotypic alterations of HeLa cells using time-of-flight secondary ion mass spectrometry. *Anal. Chem., 88*, 12196–12203.

Ju-Duo, W. A. N. G., Jia-Feng, S. O. N. G., Chang, L. I., Xin-Hua, D. A. I., Xiang, F. A. N. G., Xiao-Yun, G. O. N. G., & Zi-Hong, Y. E., (2020). Recent Advances in Single Cell Analysis Methods Based on Mass Spectrometry. *Chinese J. Anal. Chem., 48*, 969–980.

Komatsu, M., Murayama, Y., & Hashimoto, H., (2008). Protein fragment imaging using inkjet printing digestion technique. *Appl. Surf. Sci., 255*, 1162–1164.

Kompauer, M., Heiles, S., & Spengler, B., (2017). Atmospheric pressure MALDI mass spectrometry imaging of tissues and cells at 1.4-μm lateral resolution. *Nat. Methods, 14*, 90–96.

Lanni, E. J., Rubakhin, S. S., & Sweedler, J. V., (2012). Mass spectrometry imaging and profiling of single cells. *Journal of Proteomics, 75*, 5036–5051.

Li, H., Smith, B. K., Shrestha, B., Márk, L., & Vertes, A., (2015). Automated cell-by-cell tissue imaging and single-cell analysis for targeted morphologies by laser ablation electrospray ionization mass spectrometry. In: *Mass Spectrometry Imaging of Small Molecules* (pp. 117–127). Humana Press, New York, NY.

Li, L., Garden, R. W., & Sweedler, J. V., (2000). Single-cell MALDI: A new tool for direct peptide profiling. *Trends Biotechnol., 18*, 151–160.

Liu, S., Zheng, W., Wu, K., Lin, Y., Jia, F., Zhang, Y., & Wang, F., (2017). Correlated mass spectrometry and confocal microscopy imaging verifies the dual-targeting action of an organoruthenium anticancer complex. *Chem. Commun., 53*, 4136–4139.

Lombard-Banek, C., Moody, S. A., & Nemes, P., (2016). Single-cell mass spectrometry for discovery proteomics: Quantifying translational cell heterogeneity in the 16-cell frog (Xenopus) embryo. *Angew. Chem., 128*, 2500–2504.

Managh, A. J., Edwards, S. L., Bushell, A., Wood, K. J., Geissler, E. K., Hutchinson, J. A., & Sharp, B. L., (2013). Single-cell tracking of gadolinium labeled CD4+ T-cells by laser ablation inductively coupled plasma mass spectrometry. *Anal. Chem., 85*, 10627–10634.

Monroe, E. B., Jurchen, J. C., Lee, J., Rubakhin, S. S., & Sweedler, J. V., (2005). Vitamin E imaging and localization in the neuronal membrane. *J. Am. Chem. Soc., 127*, 12152–12153.

Nakashima, T., Wada, H., Morita, S., Erra-Balsells, R., Hiraoka, K., & Nonami, H., (2016). Single-cell metabolite profiling of stalk and glandular cells of intact trichomes with internal electrode capillary pressure probe electrospray ionization mass spectrometry. *Anal. Chem., 88*, 3049–3057.

Nisar, M. S., & Zhao, X., (2020). High resolution mass spectrometry for single cell analysis. *Int. J. Mass Spectrom., 450*, 116302.

Oomen, P. E., Aref, M. A., Kaya, I., Phan, N. T., & Ewing, A. G., (2018). Chemical analysis of single cells. *Anal. Chem., 91*, 588–621.

Ostrowski, S. G., Van, B. C. T., Winograd, N., & Ewing, A. G., (2004). Mass spectrometric imaging of highly curved membranes during tetrahymena mating. *Science, 305*, 71–73.

Pareek, V., Tian, H., Winograd, N., & Benkovic, S. J., (2020). Metabolomics and mass spectrometry imaging reveal channeled de novo purine synthesis in cells. *Science, 368*, 283–290.

Passarelli, M. K., Pirkl, A., Moellers, R., Grinfeld, D., Kollmer, F., Havelund, R., Newman, C. F., et al., (2017). The 3D OrbiSIMS—label-free metabolic imaging with subcellular lateral resolution and high mass-resolving power. *Nat. Methods, 14*, 1175–1183.

Patel, D. K., Dutta, S. D., & Lim, K. T., (2020). Mass spectrometry for single-cell analysis. *Handbook of Single Cell Technologies* (pp. 1–17). Springer Nature.

Redeker, V., Toullec, J. Y., Vinh, J., Rossier, J., & Soyez, D., (1998). Combination of peptide profiling by matrix-assisted laser desorption/ionization time-of-flight mass spectrometry and immunodetection on single glands or cells. *Anal. Chem., 70*, 1805–1811.

Ščupáková, K., Dewez, F., Walch, A. K., Heeren, R. M. A., & Balluff, B., (2020). Morphometric cell classification for single-cell MALDI-MSI. *Angew. Chem. Int. Ed., 59*, 17447–17450.

Sheng, L., Cai, L., Wang, J., Li, Z., Mo, Y., Zhang, S., & Chen, H. Y., (2017). Simultaneous imaging of newly synthesized proteins and lipids in single-cell by TOF-SIMS. *Int. J. Mass Spectrom., 421*, 238–244.

Spitzer, M. H., & Nolan, G. P., (2016). Mass cytometry: Single cells, many features. *Cell, 165*, 780–791.

Stopka, S. A., Khattar, R., Agtuca, B. J., Anderton, C. R., Paša-Tolić, L., Stacey, G., & Vertes, A., (2018). Metabolic noise and distinct subpopulations observed by single-cell LAESI mass spectrometry of plant cells *in situ*. *Front. Plant Sci., 9*, 1646.

Svatos, A., (2011). Single-cell metabolomics comes of age: New developments in mass spectrometry profiling and imaging. *Anal. Chem., 83*, 5037–5044.

Tian, H., Six, D. A., Krucker, T., Leeds, J. A., & Winograd, N., (2017). Subcellular chemical imaging of antibiotics in single bacteria using C60-secondary ion mass spectrometry. *Anal. Chem., 89*, 5050–5057.

Van, V. P. A., Jimenez, C. R., Li, K. W., Wildering, W. C., Geraerts, W. P. M., Tjaden, U. R., & Van, D. G. J., (1993). Direct peptide profiling of single neurons by matrix-assisted laser desorption–ionization mass spectrometry. *Organic Mass Spectrometry, 28*, 1542–1546.

Wang, J., Wang, Z., Liu, F., Cai, L., Pan, J. B., Li, Z., & Mo, Y., (2018). Vacuum ultraviolet laser desorption/ionization mass spectrometry imaging of single cells with submicron craters. *Anal. Chem., 90*, 10009–10015.

Xu, M., Pan, R., Zhu, Y., Jiang, D., & Chen, H. Y., (2019). Molecular profiling of single axons and dendrites in living neurons using electrosyringe-assisted electrospray mass spectrometry. *Analyst., 144*, 954–960.

Yin, R., Prabhakaran, V., & Laskin, J., (2018). Quantitative extraction and mass spectrometry analysis at a single-cell level. *Anal. Chem., 90*, 7937–7945.

Zhang, X., Wei, Z., Gong, X., Si, X., Zhao, Y., Yang, C., Zhang, S., & Zhang, X., (2016). Integrated droplet based microextraction with ESI-MS for removal of matrix interference in single-cell analysis. *Sci. Rep., 6*, 24730–24739.

Zhao, J. B., Zhang, F., & Guo, Y. L., (2019). Quantitative analysis of metabolites at the single-cell level by hydrogen flame desorption ionization mass spectrometry. *Anal. Chem., 91*, 2752–2758.

CHAPTER 10

Discoveries of HRMS Beyond the Globe: Application in Space and Planetary Science

AKHILA NAIR,[1,2] JÓZEF T. HAPONIUK,[1] and SREERAJ GOPI[1]

[1]*Department of Chemistry, Gdansk University of Technology, Gdansk, Poland, E-mail: sreerajgopi@yahoo.com (S. Gopi)*

[2]*R&D Center, Aurea Biolabs (P) Ltd., Kolenchery, Cochin – 682311, Ernakulam, Kerala, India*

ABSTRACT

Space exploration is an upcoming area, and many researchers, in order to broaden their horizon about spatial atmosphere, have created more precise and advanced instrumental techniques, especially high resolution mass spectrometers. This chapter aims to highlight the scope of HRMS in space and planetary science by discussing its capabilities in the identification and characterization of complex organic compounds, gases, aerosols that surround various planets, meteorites, and comets. Besides, this chapter provides a comprehensive outlook of new in-house instruments hyphenated with HRMS to increase its efficiency in resolving complex matrixes of space.

High-Resolution Mass Spectrometry and Its Diverse Applications: Cutting-Edge Techniques and Instrumentation. Sreeraj Gopi, PhD, Sabu Thomas, PhD, Augustine Amalraj, PhD, & Shintu Jude, MSc (Editors)

10.1 INTRODUCTION

Space and planetary science have been a field of interest around the globe of yore. Scientists have gone beyond the earth into other planets of the solar system. Several discoveries of the past have broadened the knowledge of people about the different planets, their atmosphere, moons, and magnetospheres. Many interesting and astounding facts have been brought into limelight such as planet of Mars were wetter and warmer as well as meteorites from it had propelled onto the earth several decades later, the surface of Venus is very hot and has aerosols as well as sulfuric acid in its atmosphere, Asteroids, and Comets can fusillade the planets and their topography and composition, planets orbiting other stars, presence of black holes and so on. This expedition of new exploration on space and planets require multidisciplinary approaches (Ocampo, 1998).

The need for mass spectrometers became necessary which could provide quick and reliable data on meteoroids and chemical composition of planet and small bodies of environment. Some MS were even accompanied rocket flights and probes, such as Cassini Huygens for exploring the surface and atmosphere of Titan and Mars (Nier, 2016). The chemical complexity of these environments demands a high revolving power system which found solace in high resolution mass spectrometer (HRMS) or Orbitrap mass analyzer (Briois et al., 2016).

This chapter deals with high resolution (HR) spectrometers used for the study of complex organic compounds, volatile as well as semi-volatile gases which forms a part of the extra-terrestrial planetary bodies like the environment of comets, Mars, and Titan. To illustrate Mars and Titan have similarities to earth, so exploring their environment is of great interest to scientists (Raulin et al., 2012).

10.2 DETERMINATION OF COMPLEX ORGANIC SOLIDS

Titan is the largest moon of Saturn and was discovered by Dutch astronomer Christiaan Huygens in 1655. This satellite is considered to exhibit similarities with the primitive atmospheric situations of Earth and so the knowledge around its chemistry is inquisitive as well as essential. Through ages, it became known that the atmosphere of Titan consists of nitrogen and small portions of methane (Raulin et al., 2012). Another compound that garnered interest is tholins which is composed of C-N-H molecules and can stimulate reddish properties onto the surface of trans-Neptunium bodies, centaurs, and icy satellites. Their m/z range to about macromolecules and polar solubility depends upon N abundance and range from 30–35%. Although their structure and composition are complex, their soluble fractions are analyzed by high resolution mass spectrometry (HRMS). The data explained the presence of at least one peak at every m/z junction till 800 u and the molecules detected depicted a polymeric structure (Carrasco et al., 2009; Sarker et al., 2003; Pernot et al., 2010). For the identification of major chemical components that compose tholins and studying the separation of isobaric compounds, Gautier et al. (2016) coupled HPLC with HRMS. The analysis of these tholins led the identification of seven molecules such as cyanoguanidine; melamine; 6-methyl-1,3,5-triazine-2,4-diamine; 2,4-diamino 1,3,5-triazine; 3-amino-1,2,4-triazole; 3,5-Dimethyl-1,2,4-triazole. However, certain compounds like hexamethylenetriamine (HMT) could not be detectable (Gautier et al., 2016). Similarly, HCN polymers are organic complexes and are very reactive compounds which can be easily hydrolyzed to form prebiotics molecules and amino acids. They are considered to be the main source of refractory carbonaceous components of comets and CN sources in cometary atmospheres. Hypothetically, HCN are constituents of tholins. Both their constituents possess optical properties in visible range similar to that denoted by the reddish surface of icy satellite, Titan's aerosol and so their study is important. Therefore, Vuitton et al. (2010), through a comparative study between tholins and HCN polymers through high resolution mass spectrometry (MS) and infrared spectroscopy (IR) established the fact that both have similar compositions. Moreover, HRMS, and MS/HRMS, revealed that HCN polymers are minors of tholins (Vuitton et al., 2010). HRMS also has a promising contribution in the study of complex organic matter in meteorites. HRMS coupled with Desorption electrospray ionization (DESI) along with an Orbitrap mass spectrometer (DESI/HRMS) was capable of identifying various CHN compounds such as alkylated pyridine, imidazole homologs which are normally difficult to analyze because of the presence of minerals. For this, an electrically charged spray of methanol having high spatial resolution of about 50 μm was utilized with HRMS for integrating the complex matrix of a simulated spatial distribution of a meteorite with a slight difference in the position of alkylpyridines and alkylimidazoles. It was observed that the chromatographic separation effect was a direct reflection of fluid movement in the parent body of the meteorite and distribution of compounds indicates each compounds respective source. Hence HRMS proved effective in the findings connected to meteorites (Naraoka and Hashiguchi, 2016). Following the previous statements that many interplanetary bodies like comets and asteroids have complex organic molecules which encounter detection challenges linked to very low concentration, diversity of molecules, ambiguous origins (abiotic or biotic), etc. In this context Eddhiff et al. (2018) investigated the potential of in-house 1D-LC-HRMS (combination of serially coupled columns with HRMS) and Dual LC (linking a series of trapping column with analytical reverse-phase column). In this in-house setup a 100 μg residue was produced from the photo-thermochemical process of a cometary ice analog. The 1D-LC-HRMS allowed to differentiate between varied chemical families of sugars, amino acids, oligopeptides, and nucleobases, facilitated by one single chromatographic run that neither supported chemical derivatization nor prior acid hydrolysis. Dual LC couples on-line enabled high mass compounds ranging from m/z 350 to 600 to be effortlessly pre-concentrated, separated, and detected (Eddhif, 2018).

10.3 DETERMINATION OF VOLATILE AND SEMI-VOLATILE GASES THROUGH COUPLED HR-TOFMS

There are numerous ways which contribute to the complexity of organic compounds in the atmosphere of earth, which is inclusive of plant emissions (Biogenic emission), combustion processes (biofuels, fossil fuels and biomass) and atmospheric oxidation processes. This complexity contributes to diverse chemical structures, formulas, physicochemical properties like volatility, oxidation degree, etc. Hence, their detection, quantification, and characterization become challenging. The versatility of time of flight (TOF) mass spectrometers are reflected in further study by Isaacman-VanWertz et al. (2017) for measuring the particle phase and gas phase of organic compounds. High resolution mass spectrometers have eased this challenging scenario as it is capable of detecting a huge range of moieties simultaneously, but the selectivity and ionization schemes pose selectivity problems that were proposed to be resolved by multiple combination of instruments. In support, a study was conducted on four high resolution time of flight mass spectrometers (HR-TOF-MS) with mass resolution 4,000 Dm/m where, 'Dm' is the mass difference between the isotope of interest and 'm' is the nominal mass of the isotope, time resolution from minutes to seconds involving different sampling and collection technique to study the functionality and volatility at different ranges. These newly developed HR spectrometers were Proton transfer mass spectrometer (PTR), chemical ionization mass spectrometer using iodide (I-CIMS), Chemical ionization mass spectrometer using nitrate (NO_3-CIMS), Thermal Denuder-Aeroyne Aerosol Mass spectrometers. In PTR, a 1 m PEEK polymer capillary (heated to 80 °C) was used for air sampling, having a non-heated Teflon filter for the prevention of particles to the detector. The analytes which are driven to the drift tube react with H_3O result an ion of molecular formula H^+ which helps to detect analytes with proton having higher affinity than water. In I-CIMS, sampling of air were facilitated at 1 lpm through 20 cm of 6.35 mm (1/4") O.D. stainless steel tubing reaching a chamber that is connected by a 1 meter Teflon line to the instrument. Also a 9 lpm of pure nitrogen was introduced to the sampling line for accelerating the residence time and diluting the sample from which the instrument sub-sampled 2 lpm. Methyl iodide is used to generate iodide ions with the help of a polonium source, which are reacted with the sample to form a cluster of ion-molecule that are thereafter analyzed and detected by mass spectrometers. NO_3-CIMS, utilizes 10 cm stainless steel tubing for air sampling at 1 lpm followed by dilution as a laminar sheath by 9 lpm pure N_2 to prevent loss to the walls. This reaction between the sample and NO_3 is carried out by X-ray ionization of nitric acid to form ion-molecule cluster, which are analyzed and detected by MS to result in a molecular formula with NO_3^+. TD-AMS (Thermodenuder aerosol mass spectrometer) were used for particle properties and not mass. Similarly, scanning mobility particle sizer (SMPS) and I-CIMS with filter inlet for gas and aerosols (FIGAERO) were helpful for the measurement of particle-phase compounds. This strategy utilized multiple chemical spaces like oxygen number vs carbon number, oxidation state of carbon vs volatility was filled with ions from these instruments and particle-phase α-pinene as well as gas were used as test mixture of organic compounds. Successful identification of overlaps (less than 10% of the measured mass) was observed within each phase and in between a particular ion and other regions in parameter space amidst particle phase and gas measurements but there were gaps observed in covering the parameters. The stated framework of the instrument was capable of covering a variety of moieties. However, there are compounds present in the atmosphere which are not much known exhibit an uncertainty in the identification of all samples (Isaacman-VanWertz et al., 2017).

Similarly, there are numerous in-house high resolution mass spectrometers built for space missions such as ROSINA/RTOF (Hässig et al., 2015). Another in-house instrument demonstrated by Hässig and coworkers was multibounce time of flight (MBTOF) for measurement of ions and neutrals to integrate the composition of volatile and semivolatile gases (Figure 10.1) (Hässig et al., 2015). This instrument with the advantage of the high mass resolution and high mass range helped to identify as well as quantify unknown species. MBTOF prototype utilizes a linear flight path with dual-lens stack. The ions get bounced in between the mirrors for a particular time period in order to increase the resolution as well as flight time and the number of bounces help tune solution of the instrument. The instrument was built for the optimization of focal point of ion optics and multi bounce passage of ion optics, however, it was also operated at linear mode to showcase minimum capabilities where the mass resolution was equivalent to a quadrupole mass spectrometer. The sensitivity in linear mode was found to be 10^{-16} to 10^{-14} cm^{-3}, the dynamic range (ratio between peak and noise) was found to be 10^2–10^3, the resolution for noble gas measurement in linear mode was found to be 100. Hence, MBTOF found effective in deep space, planetary, and terrestrial measurement (Hässig et al., 2015). Similar instrument was useful for the investigation of Strontium isotopes and Martian meteorites (Anderson et al., 2012, 2015).

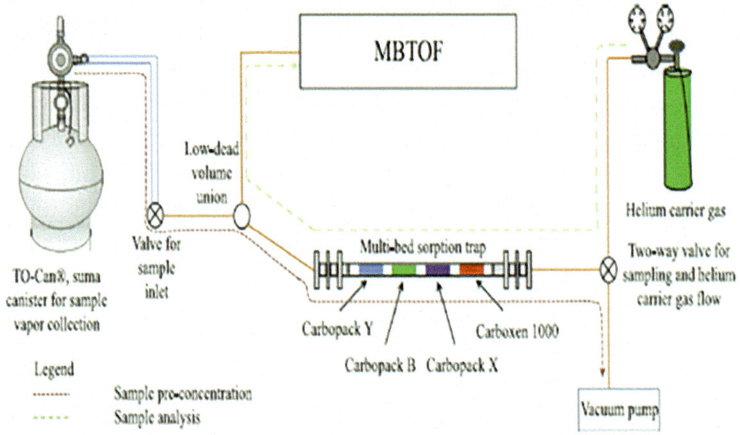

FIGURE 10.1 Diagram of the sample preparation preconcentration setup provided by the multi-bed sorption trap.

Source: Reprinted with permission from Hässig, 2016. © Elsevier.

10.4 NEW GENERATION HYPHENATED TOFMS FOR THE DETERMINATION OF ENVIRONMENTAL CONDITIONS

NASA has been instrumental in using space flight mass spectrometers for the satellites of the solar system as well as *in situ* probes for identifying and characterization of the surface and atmosphere of planets such as Enceladus, Venus, Mars, Jupiter, Saturn, and its moon Titan. For Cassini Huygens mission in 2005 to Saturn and its moon Titan, Huygens probe QMS mass spectrometer proved its potential in revealing the atmospheric composition and isotope ratios for elemental abundance. These observations led to the fact that the moon Titan is chemically active and have a resemblance to Earth's primitive atmosphere and other important revelation of the solar system. However, on the one hand, spaceflight instrument encounters various challenges in context of extremes of environment that elevate their complexity, such as temperature ranges from 243 to 600°C, caustic chemical vapors, radiation exposure, high atmospheric pressure (AP) and on the other hand science payload size, power consumption, reduce mass and funding adds worsen the situation. Hence, continuous efforts of NASA Goddard Space Flight center resulted *in situ* laboratory-based experiments involving miniature hybrid high resolving power instrument such as a micro electro mechanical system (MEMS) and nano electromechanical system (NEMS), which creates measurement reductancy as well as maintain the measurement performance providing more scientific data permission. Another new generation high resolution mass spectrometers used for environmental bodies and resolve the complexity of the chemical composition of planetary bodies is 'Cosmorbitrap' reported by Briois in 2016 (Figure 10.2). This *in situ* space exploration instrument is beneficial in the context of analyzing mass ranging in between 9 μm and 208 μm, mass resolving power of 474,000 at m/z 9 which is adjustable and escalate proportionally with transient length in the zero collision limit, minimal electric voltage required, which help measurement of negative and positive ion by switching the polarity of the electrode and reduced size of about 40 mm in diameter, 60 mm long, 250 g for a D30 cell. This prototype instrument demonstrated good agreement with commercial instruments and exhibits the capability of distinguishing between isobaric interferences, the relative isotopic abundance was below 5% and mass measurement accuracy was below 15 ppm. The background pressure is a critical parameter for any *in situ* atmospheric bodies' discovery. In terms of Cosmorbitrap, the upper limit for Nitrogen (N_2) was about 10^{-8} mbar which revealed its potential in space exploration (Briois, 2016).

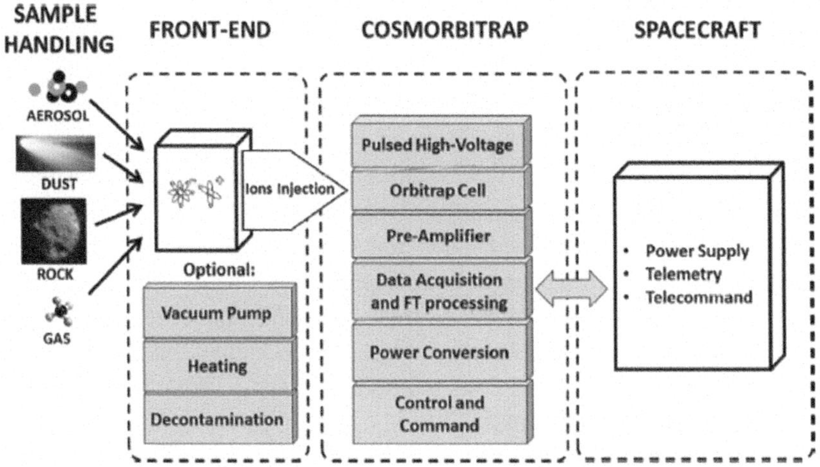

FIGURE 10.2 Elements of a Cosmorbitrap-based UHRMS instrument.
Source: Reprinted with permission from Briois et al., 2016. © Elsevier.

10.4.1 PRINCIPLE AND INSTRUMENTATION OF MEMS, NEMS, COSMORBITRAP

The basic principle behind MEMS is the ionization of gas sample and neutral spices utilizes MEMS silicon micro leaks and microvalves to maintain constant pressure in the ultra-high vacuum sector of the instrument, which has an ion lens assembly consisting of nine vertically aligned parallel ion optics lens plates which are held by two supporting rails and surface electrode deposition develops energy regions for ion acceleration and focusing. The principle behind NEMS carbon nano-field emission electron gun is that the gaseous samples of planetary soils and their moon that are subjected to pyrolysis processes and gas leak contributes to gaseous neutral species which are ionized through electron ionization (EI) inside ion source. Improved MS instrument results as the field emission electron source utilized favor lack of heated filament, lower power consumption, low operating temperature and development of micro and nanoscales. This reduced thermal energy results in low side reactions, fractionation, and contamination of ion source in the ionization procedure. The targeted performance demonstrated by it is 100 μA with electric current 70 to 100 volts in the ionization region of the ion source. Hence, these miniature TOFMS proves fruitful for planetary and terrestrial investigation (Romans, 2008).

Cosmorbitrap consists of four electrodes, namely central inner spindle shape electrode, deflection electrode and two external electrodes. The principle behind this instrument is the generation of pulsed ions with the help of high voltage central electrode and outer electrode that get polarized to form a Quadro logarithmic electric field inside the cell and deflection electrode (polarized at 0–350 V) works to optimize the injection of ions and recoup with the internal field aberration formed due to injection aperture. The transient increase in voltage of central electrode promotes fluctuation in field strength followed by an electrodynamic squeezing process where the ions trap around the central electrode by keeping a low level of high voltage of about 25,000 v for positive ions before injection that gradually get accelerated to 35,000 v during injection taking about 100–200 μs. The trapped ion axially oscillates with frequencies 0.1–7 MHz for m/z in the range of 1–4,000 (Briois, 2016).

10.5 CONCLUSION

HRMS or TOF-MS has proven instrumental in facilitating space and planetary exploration. The complex organic matter, volatile, and semi-volatile gases in the atmosphere of planets, comets, and meteorites have been effectively studied by these high resolution mass spectrometers. A coherent and comprehensive study is facilitated by these instruments for Tholins and HCN polymers which form the part of complex organic matter. Besides, the miniature laboratory-scale hybrid mass spectrometers such as MBTOF, cosmorbitrap, MEMS, and NEMS have the parallel potential to commercial instruments in terms of resolving power and are promising to resolve the complex organic matters of the atmosphere and other environmental conditions around planets, comets in deep space.

ACKNOWLEDGMENT

The authors of this chapter take this chance as an esteemed privilege to show their gratitude to the management of Aurea Biolabs (P) Ltd., Cochin, India, who were an invariable source of support to put pen to paper and encouraged us during the entire course of this chapter. We also grab this golden opportunity to express deep appreciation to our colleagues for their valuable support that accelerated the successful completion of this task.

DECLARATION OF INTEREST

The authors profess no declaration of interest in this present chapter.

KEYWORDS

- **Cosmorbitrap**
- **high-resolution mass spectrometer**
- **multibounce time of flight**
- **nanoelectromechanical system**
- **planetary science**
- **space flight mass spectrometers**
- **thermodenuder aerosol mass spectrometer**

REFERENCES

Anderson, F. S., Levine, J., & Whitaker, T. J., (2015). *Rapid Commun. Mass Spectrom., 29*, 191–204.

Anderson, F. S., Nowicki, K., Whitaker, T., Mahoney, J., Young, D., Miller, G., Waite, H., et al., (2012). *Aerosp. Conf. IEEE,* 1–18.

Briois, C., Thissen, R., Thirkell, L., Aradj, K., Bouabdellah, A., & Amirouche, B. A., (2016). Orbitrap mass analyzer for *in situ* characterization of planetary environments: Performance evaluation of a laboratory prototype. *Planet. Space Sci., 131*, 33–45.

Carrasco, N., Schmitz-Afonso, I., Bonnet, J. Y., Quirico, E., Thissen, R., & Dutuit, O., (2009). *J. Phys. Chem. A, 113*, 11195–11203.

Eddhif, B., Allavena, A., Liu, S., Ribette, T., Mrad, N. A., & Chiavassa, T., (2018). Development of liquid chromatography high resolution mass spectrometry strategies for the screening of complex organic matter: Application to astrophysical simulated materials. *Tal., 179*, 238–245.

Gautier, T., Schmitz-Afonso, I., Touboul, D., Szopa, C., Buch, A., & Carrasco, N., (2016). Development of HPLC-Orbitrap method for identification of N-bearing molecules in complex organic material relevant to planetary environments. *Icarus, 275*, 259–266.

Hässig, M., Libardoni, M., Mandt, K., Miller, G., & Blase, R., (2015). Performance evaluation of a prototype multi-bounce time-of-flight mass spectrometer in linear mode and applications in space science. *Planet. Space Sci., 117*, 436–443.

Isaacman-VanWertz, G., Massoli, P., O'Brien, R. E., Nowak, J. B., Canagaratna, M. R., & Jayne, J. T., (2017). Using advanced mass spectrometry techniques to fully characterize atmospheric organic carbon: Current capabilities and remaining gaps. *Faraday Discuss. RSC, 200*, 579–598.

Naraoka, H., & Hashiguchi, M., (2018). *In situ* organic compound analysis on a meteorite surface by desorption electrospray ionization coupled with an Orbitrap mass spectrometer. *Rapid Commun. Mass Spectrom., 32*, 959–964.

Nier, K. A., (2016). The development of mass spectrometry in the earth and planetary sciences. In: Gross, M. L., & Caprioli, R. M., (eds.), *The Encyclopedia of Mass Spectrometry, Historical Perspectives, Part A: The Development of Mass Spectrometry* (Vol. 9, pp. 222–230). Elsevier Ltd, United States. https://doi.org/10.1016/C2009-0-05659-1.

Ocampo, A., Friedman, L., & Logsdon, J., (1998). Why space science and exploration benefit everyone. *Space Policy, 14*, 137–143.

Pernot, P., Carrasco, N., Thissen, R., & Schmitz-Afonso, I., (2010). *Anal. Chem., 82*, 1371–1380.

Raulin, F., Brasse, C., Poch, O., & Coll, P., (2012). Prebiotic-like chemistry on titan. *Chem. Soc. Rev., 41*, 5380–5393.

Roman, P. A., Getty, S. A., Herrero, F. A., Hu, R., Jones, H. H., Kahle, D., King, T. T., & Mahaffy, P., (2008). A miniature MEMS and NEMS enabled time-of-flight mass spectrometer for investigations in planetary science. In: *Micro (MEMS) and Nanotechnologies for Space, Defense, and Security II.* International Society for Optics and Photonics. 6959, 69590G.

Sarker, N., Somogyi, A., Lunine, J. I., & Smith, M. A., (2003). *Astrobiology, 3*, 719–726.

Vuitton, V., Bonnet, J. Y., Frisari, M., Thissen, R., & Quirico, E., (2010). Very high-resolution mass spectrometry of HCN polymers and tholins. *Faraday Discuss., 147*, 495–508. RSC.

CHAPTER 11

Application of High-Resolution Mass Spectrometry in the Analysis of Fossil Fuels

XUEMING DONG[1,2] and XING FAN[1,3]

[1]*College of Chemical and Biological Engineering, Shandong University of Science and Technology, Qingdao, Shandong – 266590, China*

[2]*Department of Chemistry, Purdue University, Brown Building, 560 Oval Drive, West Lafayette, Indiana – 47907, United States*

[3]*Key Laboratory of Coal Clean Conversion and Chemical Engineering Process, Xinjiang Uygur Autonomous Region, College of Chemistry and Chemical Engineering, Xinjiang University, Urumqi – 830046, China, E-mail: fanxing@cumt.edu.cn*

ABSTRACT

Fossil fuels are inevitable part of everyday life, and so their analysis find a prime position in the system. Like many other natural moieties, the fossil fuels, especially the heavy crude oils and coals are composed of complex matrices. Though the conventional methods provide valuable information regarding the structure and characters of the components, high resolution spectrometric techniques can be aided for the rapid determination of tens of thousands of compounds simultaneously and to gather refined as well as in-depth information on the structural properties.

High-Resolution Mass Spectrometry and Its Diverse Applications: Cutting-Edge Techniques and Instrumentation. Sreeraj Gopi, PhD, Sabu Thomas, PhD, Augustine Amalraj, PhD, & Shintu Jude, MSc (Editors)

11.1 INTRODUCTION

Fossil fuels, including both petroleum and coal, are the key sources of energy and material for the modern industrial world. With the recent depletion of "sweet" light crude oils, the global energy market increasingly relies on other options such as heavy crude oils and coals (Rana et al., 2007). Compared to the light crude oils, heavy crude oils and coals are normally harder to upgrade, which inevitably requires a better understanding of the structures of the heavy crude oils and coals (Mullins et al., 2007; Mamin et al., 2016). Unfortunately, heavy crude oils and coals are hard to analyze because they are composed of hundreds of thousands of compounds and are among the most complex organic mixtures on earth (Marshall and Rodgers, 2004). To differentiate from conventionally regarded complex mixtures such as proteome and metabolome, fossil fuels are regarded as ultra-complex mixtures in this chapter.

It needs to be noted that a wide range of analytical methods with unique advantages and drawbacks have been employed to study fossil fuels. In general, conventional spectroscopic methods such as Fourier transform infrared spectroscopy (FTIR), fluorescence depolarization, nuclear magnetic resonance (NMR), Raman spectroscopy, and X-ray diffraction mainly provide bulk information about fossil fuels (Melendez et al., 2012; Chen et al., 2012; Badre et al., 2006; Ghosh et al., 2007; Hasan et al., 1983; Evdokimov et al., 2003; Rausa et al., 1989; Jing et al., 2019; Dias et al., 2015; Dearing et al., 2011; Chung et al., 2000; Asher et al., 1984; Yen et al., 1961; Lu et al., 2001). Electron microscopy (EM) has been widely accepted as a valuable tool to study macromolecular structures of fossil fuels (Pérez-Hernández et al., 2003; Van Geet et al., 2001; Ye et al., 2013). These methods provide a good measurement of the bulk properties of fossil fuel samples, but lack the ability to determine the molecular structures of compounds. On the other hand, impressively, the structures of several molecules derived from both coals and crude oils were determined by atomic force microscopy (AFM) (Schuler et al., 2017; Hsu et al., 2011). However, such studies only determined the molecular structures for an extremely small portion of the molecules in a sample. Consequently, making statistically sound conclusions based on these observations would be challenging.

High-resolution mass spectrometry (HRMS) is an excellent analytical method capable of rapidly providing elemental compositions of tens of thousands of ionized compounds (Hsu et al., 2011). Although HRMS is usually unable to determine the exact structure of the molecules in a fossil fuel sample, high-resolution tandem mass spectrometry (MS/MS) reveals important structural information about the compounds that is useful for the later refining of the fossil fuel samples (Quann et al., 1992; Qian et al., 2012). In this regard, MS analysis of fossil fuels offers in-depth structural information of a large portion of the ionized sample (Figure 11.1). Overall, HRMS provides a good balance between the details and the speed of the analysis.

Nevertheless, it must be noted that MS standing alone is not a quantitative method. This is due to that the ionization efficiency is quite different for molecules with different structures, and various compounds compete for the charge during the ionization process. This problem can be addressed by using several methods such as stable-isotope standards for some of the conventional complex mixtures. However, currently there are few established quantitative methods for ultra-complex mixtures such as heavy petroleum.

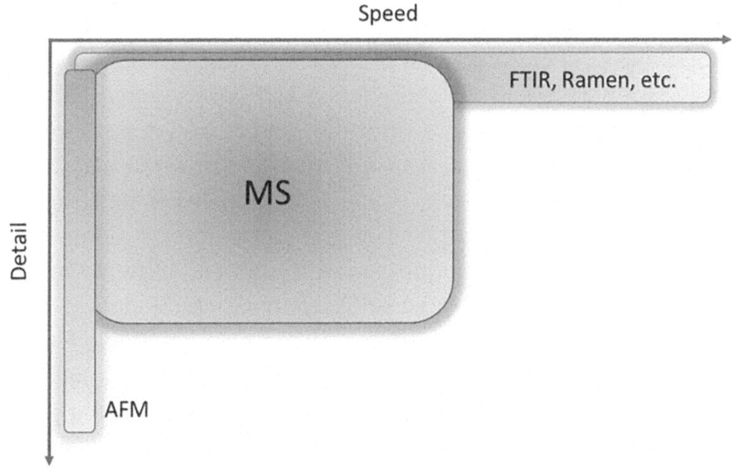

FIGURE 11.1 Comparison of different analytical techniques in terms of analytical speed and detail.

11.2 PETROLEOMICS

High-resolution and ultrahigh-resolution MS is the key and prerequisite to the relatively new field "petroleomics," which was first introduced by

Marshall and Rodgers (2004). Petroleomics aims to identify all of the compounds in the fossil fuel samples and find the correlations between the chemical composition and physio-chemical properties of the fossil fuels (Marshall and Rodgers, 2004). In the following part of this section, the basic concepts and techniques related to petroleomics will be introduced, and the later sections will describe the petroleomics studies of heavy crude oils and coals.

The first and probably most important concept in petroleomics is mass resolution, or mass resolving power, of a mass spectrometer. In this chapter, it is defined as the following, although the definition might vary in other fields:

$$Mass\ Resolution = \frac{m}{\Delta m_{50\%}} \tag{11.1}$$

where; m is the mass of the ions; and $\Delta m_{50\%}$ is the width of the mass spectral peak at half of its height.

Theoretically, each chemical formula, or elemental composition, has its unique mass. Therefore, with a sufficient mass resolving power, the chemical formula of ionized compounds can be accurately determined. However, in practice, the mass resolving power of a mass spectrometer might not be sufficient. Thus, it is important to know the mass resolving power of a mass spectrometer when analyzing fossil fuel samples. In general, the majority of the Fourier transform ion cyclotron (FTICR) mass spectrometers, and some of the latest high-end Orbitrap mass spectrometers, are very likely to have sufficient mass resolving power for the unambiguous determination of elemental composition alone. Other mass spectrometers, such as time-of-flight (TOF) and conventional Orbitrap mass spectrometers, might not have sufficient mass resolving power for the unambiguous determination of elemental composition. In general, in order to claim unambiguous elemental composition determination, the root-mean-square deviation (RMS) of the error between the mass of an assigned elemental composition and the measured mass of the corresponding mass spectral peak needs to be less than 300 ppb (Kim et al., 2006; Stenson et al., 2003).

In addition, it also needs to be emphasized that the required mass resolving power also depends on the nature of the sample. For example, a mixture composed of only hydrocarbons (C and H only) requires a much less resolving power to unambiguously assign the elemental composition than a heavy crude oil that might contain C, H, N, O, and S elements. Therefore, the exact requirement for the mass resolving power should be determined in a case-by-case scenario.

A correctly assigned chemical formula can be used to obtain a double bond equivalence (DBE) value, which is also referred to as ring and double bond equivalent (RDBE) value. The DBE value represents the degree of unsaturation of a compound and can be calculated based on the following equation:

$$DBE = C - \frac{H}{2} - \frac{X}{2} + \frac{N}{2} + 1 \tag{11.2}$$

where; C, H, X, and N are the number of carbon atom, hydrogen atom, halogen atom, and nitrogen atom, respectively, in the compound.

The number of carbon atoms, DBE value, and the intensity of the mass spectral peaks can be used to establish a DBE plot. The DBE plot is frequently used to represent fossil fuel samples in alternative to the conventional mass spectrum. An example of the DBE plot and the mass spectrum is illustrated in Figure 11.2 (Fan et al., 2018).

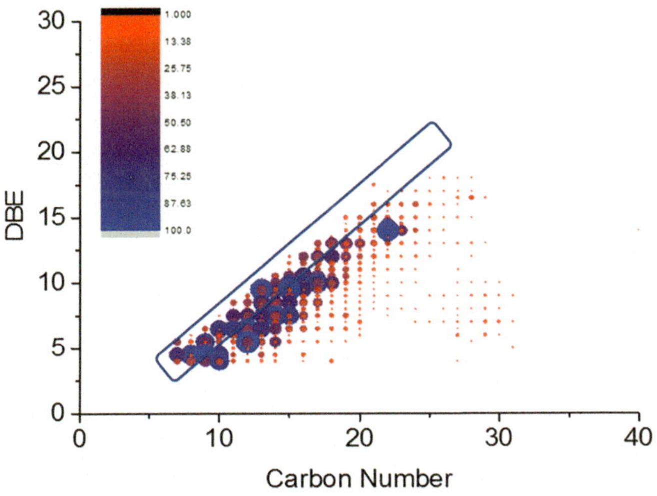

FIGURE 11.2 A DBE plot derived from the cyclohexane thermal dissolution product of a lignite sample.

Source: Reprinted with permission from Fan et al., 2018. © Elsevier.

Compared to a conventional high-resolution mass spectrum (Figure 11.3), which shows the data in two dimensions, mass to charge ratio (m/z) and relative intensity, the DBE plot illustrates the data in three dimensions, carbon number, DBE value, and relative intensity, and thus provides an arguably better representation of the complicate sample.

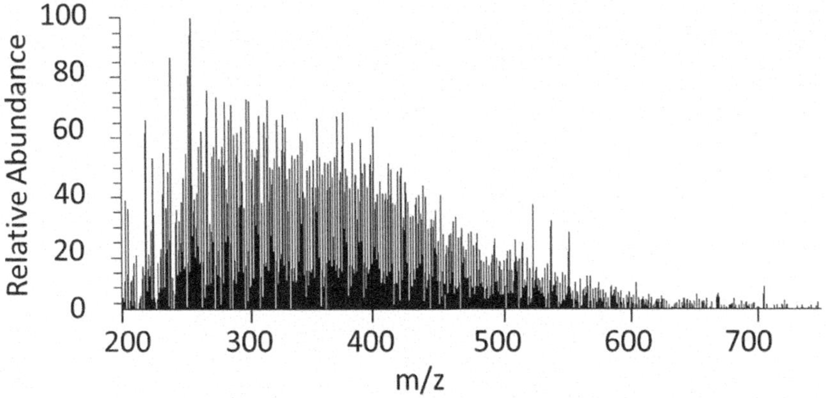

FIGURE 11.3 A high-resolution mass spectrum of an ionized asphaltene sample.

As discussed earlier, in addition to the characterization of all the compounds in fossil fuel samples, petroleomics also aims to study the correlations between the chemical properties of the fossil fuels and their physio-chemical properties. Unfortunately, until recently, publications on this subject have been limited to the total content of elements in samples, the American Petroleum Institute (API) gravity and Conradson Carbon Residue (CCR) (Corilo et al., 2016; Santos et al., 2018; Lozano et al., 2017; Dong et al., 2019). The chemometric approach is mainly hindered by the lacking of an ideal universal ionization method that ionizes different compounds with an equal efficiency and will be discussed in the next section.

11.3 IONIZATION METHODS IN MASS SPECTROMETRIC ANALYSIS OF FOSSIL FUELS

A mass spectrometer measures the m/z of ionized compounds, and the molecular weight is calculated from the mass over charge ratio. Therefore, ionization is the prerequisite for MS analysis and the development of ionization methods has been a constantly active field in MS. In 2002, the Nobel Prize in Chemistry was awarded to the developments of electrospray ionization (ESI) and matrix-assisted laser desorption ionization (MALDI) for the analysis of large biological molecules and polymers. Currently, ESI, atmospheric-pressure photoionization (APPI), and atmospheric-pressure chemical ionization (APCI) are widely available in commercial MS instruments (Fenn et al., 1989; Tanaka et al., 1988).

Historically, MS analysis of fossil fuels precedes the MS analysis of large biomolecules. ESI is one of the predominate ionization methods in the bioanalysis field and is highly efficient and specific for the ionization of polar compounds, which constitute only a small portion of the common fossil fuel samples. Consequently, the majority of the petroleum samples are not ionized and thus analyzed during ESI-MS characterization. Nevertheless, ESI is suitable for the scenario when the polar compounds are in the fossil fuel samples, such as naphthenic acids (Colati et al., 2013; Hsu et al., 2000; Mapolelo et al., 2011).

Characterization of the low polar parts that account for the majority of fossil fuel samples is traditionally achieved by ionization methods that do not specifically target the polar compounds. Field desorption/ionization (FD) can ionize multiple classes of hydrocarbons, including saturated hydrocarbons with a low level of fragmentation (Schaub et al., 2003, 2005). However, FD experiments are slow due to the need for the gradual increase of current to the FD emitter. This prohibits FD to be coupled with mainstream automated separation methods such as gas chromatography (GC) and liquid chromatography (LC). FD also suffers from a lower reproducibility compared to its peers due to the sensitivity to emitter shape and the presence of salts (Beckey and Schulten, 1975). Furthermore, from a practical point of view, the FD emitters are expensive and easily damaged during the experiment, rendering FD experiment expensive unless operated by highly experienced experts.

APPI ionizes the chromophore containing compounds, which usually include aromatic compounds and those containing conjugated double bonds existing in fossil fuel samples (Robb et al., 2000). Apparently, APPI suffers from the inability in ionizing saturated hydrocarbons without chromophores and therefore needs the assistance from dopants. In addition, upon APPI, one compound can generate both molecular ions and protonated ions. The protonated ions are very hard to resolve from the naturally occurring ^{13}C isotopic ions, unless analyzed by the ultrahigh-resolution

MS, such as the higher-end FT-ICR MS. Overall APPI provides a rapid sample analysis speed that is compatible with LC (McKenna et al., 2009; Purcell et al., 2006; da Silveira et al., 2016). In addition, APPI ionizes a much wider range of compounds in fossil fuels compared to ESI (Bae et al., 2010; Fernandez-Lima et al., 2009; Kim et al., 2011). Considering the rapid dwindling use of light crude oils, which are rich in saturated hydrocarbons, and the increasing reliance on heavy crude oils and coals that contain less saturated hydrocarbons in the energy and chemical industries, APPI is becoming a highly promising ionization method for the analysis of fossil fuels.

APCI is a MS ionization technique with a longer history, compared to ESI and APPI, but is being constantly developed (Carroll et al., 1974). APCI can rapidly ionize both polar and non-polar compounds, regardless of the presence of chromophores or aromatic groups (Byrdwell, 2001). This allows APCI to be coupled with LC for the analysis of a wider range of compounds (Wu et al., 2015). However, similar to APPI, APCI also suffers from forming both molecular ions and protonated ions from one compound. This issue can be largely addressed by using carbon disulfide, which does not provide the source of proton, as the ionization reagent for APCI (Owen et al., 2011). Unfortunately, carbon disulfide is not only highly flammable, but also highly toxic. Carbon disulfide has been associated with various forms of acute and chronic poisoning. Thus the use of carbon disulfide as the ionization reagent is mainly limited to a few research groups in academia. Another drawback of the conventional APCI approach is that it cannot ionize saturated hydrocarbons without causing significant fragmentation (Jin et al., 2016). Recently, several studies showed that it may be possible to reduce or minimize the fragmentation of hydrocarbons during APCI analysis by using specific ionization reagents to control the exothermicity of the ionization reaction (Tose et al., 2015).

Other than the above-mentioned popular ionization methods, there are other ionization techniques that have been employed to analyze fossil fuels. These methods include ambient solids analysis probe (ASAP), atmospheric pressure (AP) ionization coupled with GC, laser-induced acoustic desorption (LIAD) coupled with chemical ionization, and direct analysis in real-time (DART) (Abdelnur et al., 2008; Wu et al., 2015; Gao et al., 2011; Lobodin et al., 2015). Each of these methods has unique advantages and disadvantages. The interested readers are encouraged to reach the respective literatures.

As discussed above, each MS ionization method has its own pros and cons. At present, there is no hypothetically "ideal" ionization method that can rapidly ionize all compounds efficiently and reproducibly with an equal efficiency. Hence, for the MS analysis of fossil fuel samples, it is critically important to understand the nature of ionization methods and evaluate the appropriateness of the ionization methods for the analyte of interests. Notably, several studies employed multiple ionization methods to characterize fossil fuel samples (Headley et al., 2016; Fan et al., 2018; Yerabolu et al., 2018). Due to the extreme complexity of fossil fuel samples, a compressive description of one sample requires the incorporation of multiple ionization methods. In addition, separating or fractionating of the sample prior to MS analysis and using the appropriate ionization method or instrument for each fraction might further aid the comprehensive analysis of a sample (Yerabolu et al., 2018; Chacón-Patiño et al., 2017). However, the comprehensive characterization has a high cost, because using the optimal ionization method/instrument for the sample, as well as separation of the sample prior to analysis requires considerable investments in instrumentation and manpower. The cost-effectiveness of the method might be another factor to be considered in an industrial setting. The workflow of fractionation and analysis in one comprehensive crude oil characterization study is shown in Figures 11.4 and 11.5, respectively (Yerabolu et al., 2018).

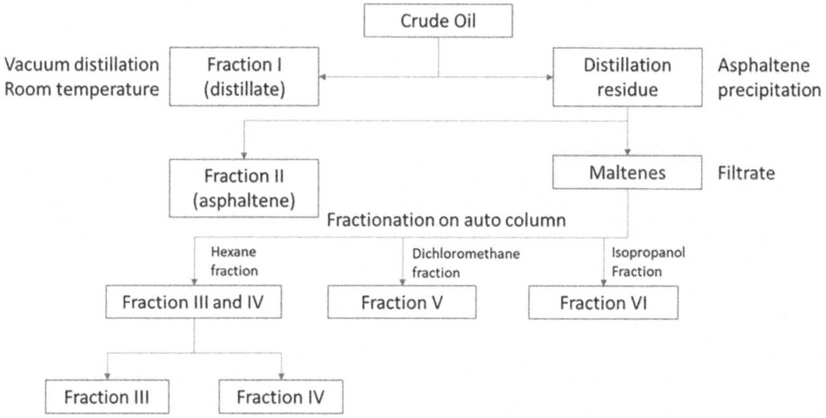

FIGURE 11.4 Distillation, precipitation, and fractionation separation methods for crude oil analysis.

Source: Reprinted with permission from Yerabolu et al., 2018. © Elsevier.

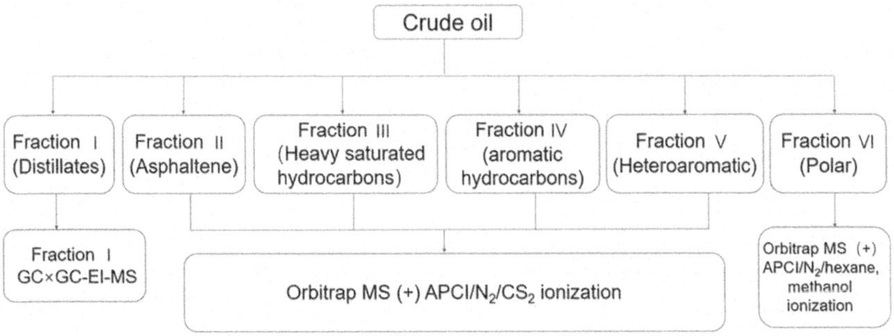

FIGURE 11.5 The instruments and ionization methods employed for the analysis of six fractions from fractionation methods as shown in **FIGURE 11.4**.

Source: Reprinted with permission from Yerabolu et al., 2018. © Elsevier.

Although the ionization method might be considered as "appropriate" by principle, the choice of suitable ionization method for a specific sample still needs to be validated and evaluated experimentally for the quantitative. One possible approach, as shown in Figure 11.6, is to statistically compare the elemental compositions derived from the HRMS analysis with those obtained from established quantitative methods such as elemental analysis based on elemental analyzer (the "actual value") (Dong et al., 2019). The ionization method that provides the closest elemental composition with the "actual value" might be considered the appropriate method for the sample of interest.

Overall, ionization method is important for a successful MS analysis of fossil fuel samples. Using the right ionization methods and the suitable prior separation methods are the key for a comprehensive and unbiased characterization of a fossil fuel sample. It must be emphasized that sometimes an ionization method that is suitable "in principle" for the analyte needs to be experimentally evaluated for the quantitative potential of measurement.

11.4 PETROLEOMIC ANALYSIS OF CRUDE OILS

As discussed in earlier sections, the term high complexity mixture is probably not enough to distinguish crude oil from other complex mixtures such as proteome. Crude oil contains at least hundreds of thousands of compounds which are a number that might be considered larger than those in the human proteome (Marshall and Rodgers, 2004). Furthermore, unlike in proteome where most of the compounds are proteins with similar chemical natures, the compounds in crude oil have different chemical natures. For a few decades, one simple but popular method is dividing the crude oil into saturate, aromatic, resin, and asphaltene (SARA) fractions (Fan et al., 2002; Jewell et al., 1972).

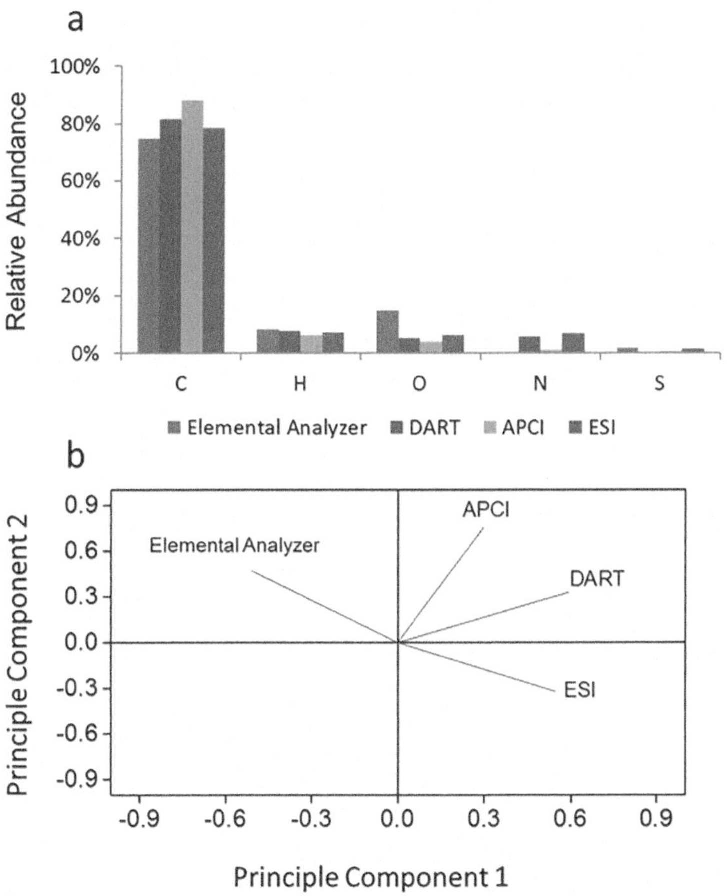

FIGURE 11.6 (a) the relative abundances of each elemental of a coal-derived liquid obtained by elemental analyzer, DART-MS, APCI-MS, and ESI-MS. (b) The principal component analysis of the elemental compositions obtained by elemental analyzer, DART-MS, APCI-MS, and ESI-MS. The element abundances based on APCI-MS are closest to those based on elemental analyzer among the three ionization methods.

Source: Reprinted with permission from Dong et al., 2019. © Elsevier.

The saturate fraction contains nonpolar compounds, including saturated hydrocarbons in different forms (linear, branched, or cyclic). The aromatic fraction contains compounds with more than one aromatic group. The resin fraction contains generally polar compounds. The asphaltene fraction is loosely defined as the compounds in crude oil that are insoluble in heptane or pentane. Generally, the saturate and aromatic fractions provide fuels and organic building blocks during the refining processes, while resin and asphaltene fractions contain corrosive, toxic, or catalyst-poisoning compounds. Asphaltene fraction is additionally prone to precipitate and clog the transportation pipelines, which significantly increases the maintenance cost (Dong et al., 2019).

Besides the academic interests that have been widely elaborated, one of the main economical reasons for the molecular-level characterization of crude oils is to provide information for the petroleum composition modeling that can be used to predict the crude oil upgrading process (Rodgers and McKenna, 2011; Jaffe et al., 2005; Ghosh and Jaffe, 2006; Christensen et al., 1999). Historically, the petroleum industry, mainly relies on the light crude oils that are easy to upgrade, and the analysis of the light crude oils has been well-established. Light crude oils with relatively low boiling points can be well characterized by using GC/MS and 2-dimensional GC-MS (GC×GC/MS) (Teng et al., 1994; Wang et al., 2003, 2005; Frysinger and Gaines, 1999; Lai et al., 1995). Due to the complexity of crude oils, GC×GC that was first introduced in 1991 provides a higher separation power than the traditional GC (Liu et al., 1991). This advantage leads to the detection of substantially larger numbers of compounds. Conventionally, electron ionization (EI) is used as the ionization method in GC/MS. EI generates highly reproducible mass spectra, which enables automatic compound identification based on database search in various libraries. Nevertheless, it must be noted that the library might not be an exhaustive collection of all possible compounds in crude oils. In addition to EI, atmospheric-pressure ionization has also been employed in multiple studies and demonstrated several potential advantages compared to EI (Portolés et al., 2010, 2012; Carrasco-Pancorbo et al., 2009).

Nowadays, as light crude oils are being rapidly depleted, the petroleum industries have to switch to heavy crude oils. The heavy crude oils have low volatility and are not suitable for GC/MS analysis. Subsequently, the analysis of heavy crude oils remains to be very challenging until the first introduction of ultra-high-resolution FTICR MS and the high-resolution Orbitrap MS (Headley et al., 2011, 2016; Qian et al., 2001; Rodgers et al., 2005; Vetere et al., 2017; Cole et al., 2013). The ultra-high-resolution MS can unambiguously determine the elemental compositions of hundreds of thousands of unknown compounds, which offer substantial advances in the heavy crude oil analysis (Marshall and Rodgers, 2004; Wittrig et al., 2017). However, it must be noted that determining the molecular formula alone is not usually considered as structural characterization. Additionally, as demonstrated by studies using ion mobility MS (IMS), several isomeric compounds were observed for one chemical formula (Fernandez-Lima et al., 2009; Lalli et al., 2015). As a result, high-resolution MS alone does not provide useful information for the mathematical modeling of a crude oil upgrading process. Therefore, MS/MS methods, which often coupled with various of separation methods, must be employed to obtain in-depth structural information.

A wide range of separation methods have been explored for the crude oil analysis. These methods range from the most traditional SARA fractionation, more recent solid-phase extraction (SPE), and the modern automated multidirectional high-performance liquid chromatography (HPLC) (Cho et al., 2012; Robson et al., 2017; Bissada et al., 2016; Putman et al., 2017). Although debates still remain, it is reasonable to argue that separation and/or fractionation of crude oils prior to analysis is at least noticeably beneficial. As discussed in the earlier sections, separation, and fractionation not only allow a better detection, but also enable the use of optimal separation, ionization, and detection methods for the fractionated samples. In the absence of prior separation, the analysis may enormously favor one type of compounds and lead to controversial conclusions. The relative abundances of single-core compounds (compound composed of a large aromatic core with peripheral alkyl chains attached) and multicore compounds (a compound with multiple aromatic cores connected via alkyl linkage) in asphaltenes have been debated for many years (Mullins et al., 2007; Strausz et al., 1992, 2002; Alvarez-Ramírez and Ruiz-Morales, 2013). There are multiple studies proposed that the multicore compounds are not abundant in asphaltenes, while others proposed that the abundance multicore compounds are dependent on the individual asphaltene samples (Nyadong et al., 2017; Wittrig et al., 2017; Chacón-Patiño et al., 2017; Podgorski et al., 2013; Jarrell et al., 2014; DeCanio et al., 1990; Karaca et al., 2009; Ha et al., 2017). Recently, two studies suggested that the single-core compounds are preferentially ionized during MS analysis, and thus the multicore compounds should be analyzed after proper separation steps (Chacón-Patiño et al., 2017).

The debate on the single-core versus multicore abundances also demonstrates the importance of MS/MS for the structural characterization of a petroleum sample. A well-elaborated example is shown in Figure 11.7, where a single-core compound and a multicore compound can have the same chemical formula (Qian et al., 2012). However, the two compounds will give drastically different products during the refining process. The single-core compound will give less valuable, highly aromatic and heavy products, while the multicore compound provide valuable low-boiling aromatic light products (Qian et al., 2012). This study brings up a very possible scenario where the determination of chemical formula alone might fail to predict the crude oil refining outcome. For two crude oils with similar composition, the crude oil containing more multicore compounds will generate lighter products than the one with more single-core compounds, and thus the former one is more valuable. This debatable rationalizes the debate on single-core versus multicore compound abundances in asphaltenes. An asphaltene composed of mostly single-core compounds, as described by the Yen-Mullins model, is deemed to be un-upgradable and non-profitable (Mullins, 2010; Mullins et al., 2012). In contrast, an asphaltene composed of abundant multicore compounds, as suggested by other studies, might be considered as a profitable feedstock. Consequently, the "heaviness" of a petroleum sample during the upgrading process, is also related to the relative abundances of single-core compounds versus multicore compounds (Chacón-Patiño et al., 2017a, b). To address this issue, MS/MS can be used to evaluate the multicore compound abundance in different petroleum samples (Nyadong et al., 2017). Upon fragmentation, ions derived from multicore compounds and single-core compounds produce fragment ions of lower and similar DBE values compared to the parent ions, respectively (Figure 11.8) (Jarrell et al., 2014; Sabbah et al., 2011; Nyadong et al., 2017). This approach has been employed to compare the abundances of the multicore compounds in petroleum samples with different origins (Figure 11.9) (Nyadong et al., 2017).

However, it is must be highlighted that the investigated ions during the MS/MS study must be statistically representative to avoid sampling errors. As demonstrated in a recent study, the conclusions regarding the abundance of multicore compounds in two asphaltene samples are different depending on the ions investigated (Figure 11.10) (Dong et al., 2019). Consequently, although not routinely carried out, it is crucial to take statistics into the consideration during the MS/MS analysis of petroleum samples. However, there is indeed a few MS/MS methods that are free from the statistical sampling issue: ion-source collision-activated dissociation (ISCAD), and multipole storage assisted dissociation (MSAD) (Dong et al., 2019; Qian et al., 2012). The former is a relatively common feature in commercial mass spectrometers.

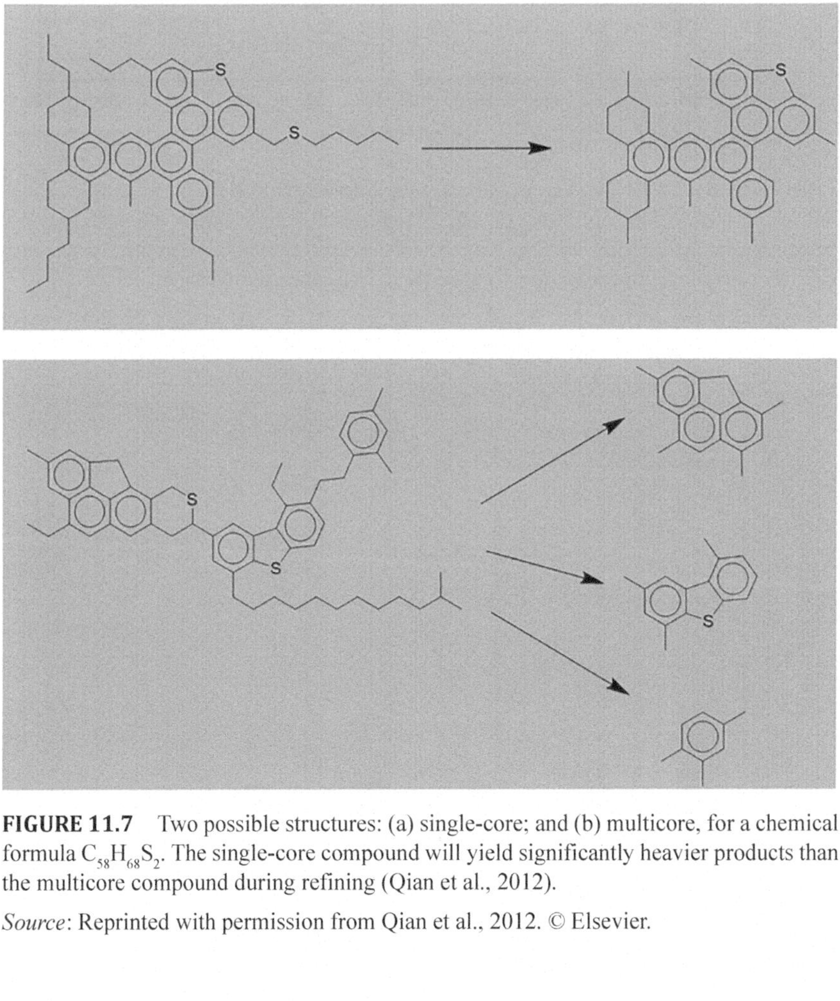

FIGURE 11.7 Two possible structures: (a) single-core; and (b) multicore, for a chemical formula $C_{58}H_{68}S_2$. The single-core compound will yield significantly heavier products than the multicore compound during refining (Qian et al., 2012).

Source: Reprinted with permission from Qian et al., 2012. © Elsevier.

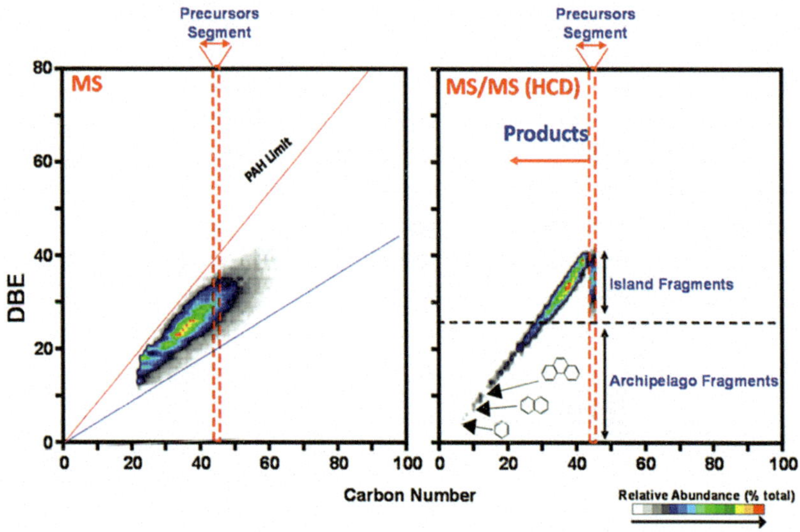

FIGURE 11.8 DBE plots of an asphaltene sample derived from (left) the full MS spectrum, and (right) the MS/MS spectrum. The ionized multicore (archipelago) compounds generate fragment ions with lower DBE values to their parent ions, while the ionized single-core (island) compounds fragment ions with the same DBE values to their parent ions.

Source: Reprinted with permission from Nyadong et al., 2018. © American Chemical Society.

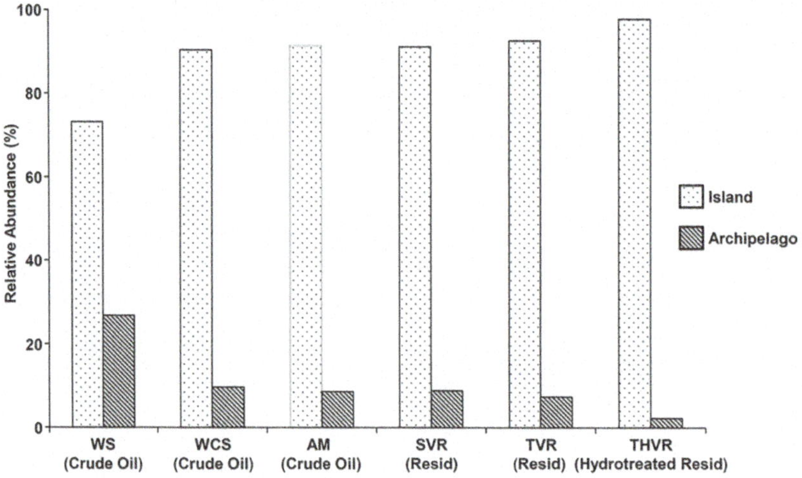

FIGURE 11.9 Relative abundances of the single-core (island) compounds versus multicore (archipelago) compounds in different petroleum samples.

Source: Reprinted with permission from Nyadong et al., 2018. © American Chemical Society.

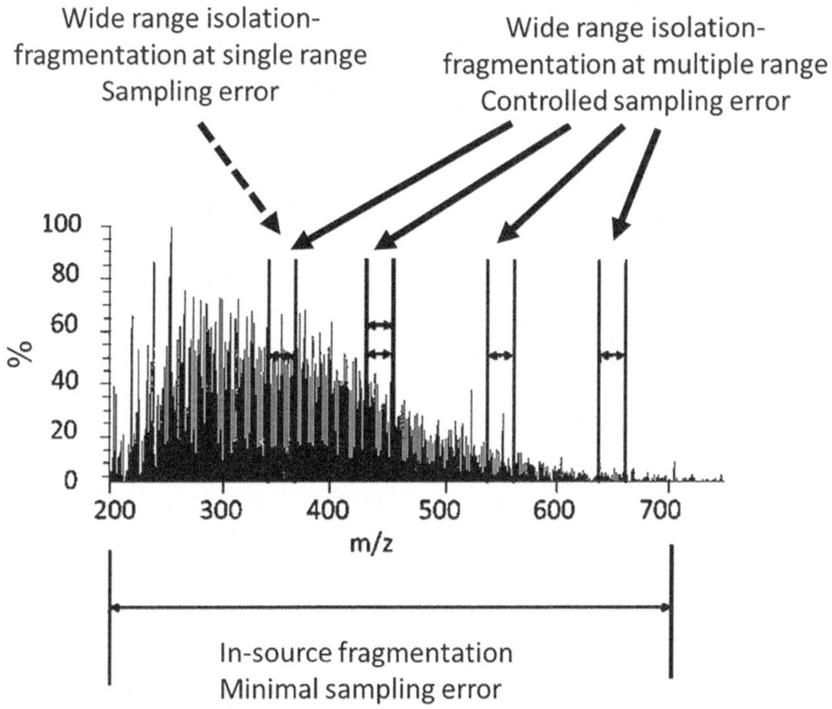

FIGURE 11.10 The correlation between the investigated ions and the statistical sampling errors during the MS/MS investigation of petroleum samples.

Source: Reprinted with permission from Dong et al., 2019. © Elsevier.

11.5 PETROLEOMICS ANALYSIS OF COALS

As a non-renewable natural resource, coal has extremely complex structures and consists of a wide range of organic species (Ashida et al., 2008; Wang et al., 2016). Coals at different ranks have various molecular compositions and structures. Even within the same rank, coal structures may be heterogeneous depending on the environmental conditions of coalification (Fan et al., 2016). Understanding the structure and composition of molecules in coals is important for effective, clean, and value-added utilization of coals. Although coal structures are under debating, it is widely agreed that low molecular weight compounds (small molecules; 500 Da or below) and macromolecules coexist in coals (Feng et al., 2012; Zubkova, 2011). Macromolecules consist of a skeletal network with small molecules scattered inside, which can be separated from the macromolecular skeleton by extraction, liquefaction, and oxidation, etc. (Xia et al., 2016; Sun et al., 2017; Yao et al., 2010).

To understand the chemical structures of coals, various analytical methods were developed to identify the organic matter (Takanohashi and Kawashima, 2002; Geng et al., 2009; Wang et al., 2018). Unlike other analytical methods providing bulk properties, high-resolution MS enables the discrimination of molecular ions with a tiny difference in m/z and provides the accurate elemental composition for complicated components in coals according to the high resolving power (Li et al., 2019).

Soluble products from thermal dissolution of two low-rank coals, SL, and HL, were analyzed by a HPLC/HRMS (Q-Exactive, Thermo-Fisher Scientific, USA) (Fan et al., 2017). The composition of identified organic species is expressed by the class of chemical formula such as O_o, O_oN_n, and O_oS_s, where subscripts o, n, and s provide the number of oxygen, nitrogen, and sulfur atoms per molecule, respectively. Most of the identified components can be assigned to a unique chemical formula on the basis of the accurate mass. Oxygen-containing compounds are the major heteroatom-containing compounds in low-rank coals (Fan et al., 2018). In both SL and HL, O_o (o = 1–6) classes have the highest relative abundances and the corresponding MS data have been processed to exhibit valuable information. In economics, the concentration ratio is defined as the percentage of market share taken up by a given number of the largest firms in an industry, which can be introduced to describe the content inequality of organic species within the same class of coal (Jovanovic, 1982). The top 5% of the compounds in the O_o classes account for 25–55% of the total relative abundances of components in the same class (shown in Figure 11.11). It is interesting that the concentration ratios of O_o classes in the HL coal are always higher than those of the SL coal. The same results are observed for the concentration ratios of O_o classes if the top 5% is extended to the top 10%. Coals with a high concentration ratio for a specific class should take priority as a feedstock for chemicals and receive more attention.

Based on the amounts of data from HRMS, multivariate statistical methods such as hierarchical cluster analysis (HCA) and principal component analysis (PCA) are used to extract valuable information through data processing. Both HCA and PCA can rapidly realize simultaneous analysis of multiple samples via reducing the complexity of data, and more similarities and/or differences in multiple samples can be presented in a single plot.

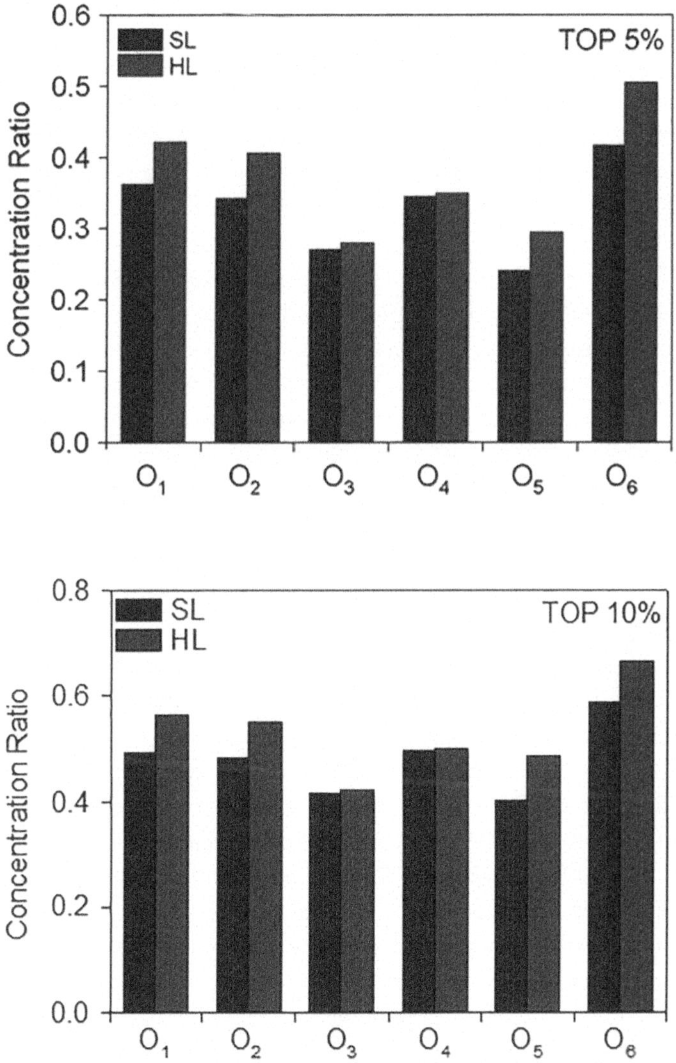

FIGURE 11.11 Distribution of 'concentration ratio' for the top 5% and 10% of compounds in O_o classes.

Source: Reprinted with permission from Fan et al., 2017. © John Wiley and Son.

HCA is often used to find "clusters" inside samples (Christensen et al., 2004; Chiaberge et al., 2013). The SL low-rank coals in the last paragraph was extracted and thermally dissolved with cyclohexane, acetone, and methanol sequentially (Yu et al., 2019). The extracts were analyzed using a HRMS (LTQ-Orbitrap XL, Thermo-Fisher Scientific, USA) and the identified species were assigned to 18 clusters (classes) according to the distribution of heteroatoms per molecule. HCA was performed to show the similarities and differences of the 18 clusters visually expressed as dendrograms and heatmaps in Figure 11.12. The numbers of O, S, and N atoms per molecule are variables determining the tree topology. The distance between S and N shown in the dendrograms on the top of heatmaps is closer compared to the distance between N/S and O. Thus the two variables (the number of S and N) can be set as a single factor, and the number of O works as a single variable factor. It is consistent with the result of element analysis where the content of oxygen is significantly higher than that of sulfur and nitrogen. The dendrograms on the left side of heatmaps display clustering paths and distances between different clusters, and the distribution of clusters according to the clustering paths is shown on the right side of heatmaps. The distance between two adjacent intersections in dendrograms is proportional to the number of identified compounds in the corresponding cluster. Such distances are obviously different, which can be ascribed to extraction conditions like temperature.

PCA aims to reduce the number of variables, extract useful information and classify samples according to principal components (PCs) (Weldegergis et al., 2011). The low-rank coal in Figure 11.11, HL, was sequentially extracted by six solvents and the six extracts were analyzed using a FTICR MS (9.4 T, Bruker, Germany) equipped with an ESI or APPI ion source (Li et al., 2019). The MS data processed by PCA are shown in Figure 11.13 as both score and loading plots, which are constructed with two PCs and can explain 70.4% of the total variance. The data sets are obviously divided into two groups. Group 1 contains the data obtained via APPI ion source under positive ion mode. Data in Group 2 are acquired by ESI-FTICR MS with both positive and negative modes. The identified hydrocarbons and heteroatom-containing compound are expressed in the loading plot.

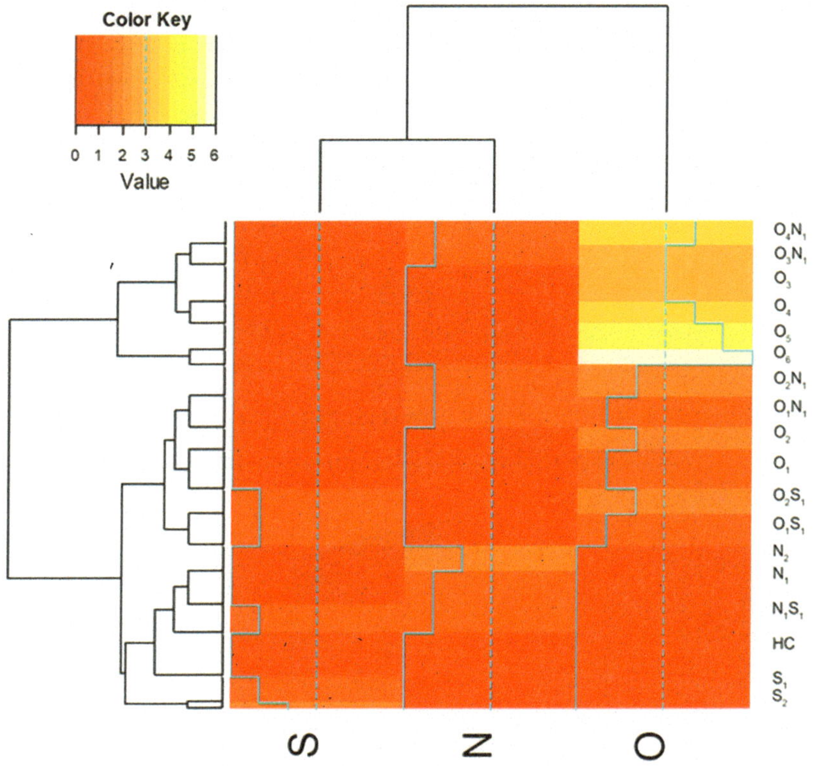

FIGURE 11.12 Dendrogram and heatmap of cyclohexane extract.

Source: Reprinted with permission from Yu et al., 2019. © Elsevier.

11.6 CONCLUSION

Methodology for the characterization of fossil fuels based on HRMS have been developed in the last 15 years, and the corresponding molecular recognition is still a challenging task. It is necessary to optimize the current MS techniques such as collision-activated dissociation, dopant-assisted ionization, and data processing to get accurate and meaningful molecular information for fossil fuels. Newly emerging MS techniques like ion mobility should also be noticed for their application in fuel analysis. Petroleomics should not be looked at as just an analytical strategy, but a methodology actively participating in fuel conversion and upgrading.

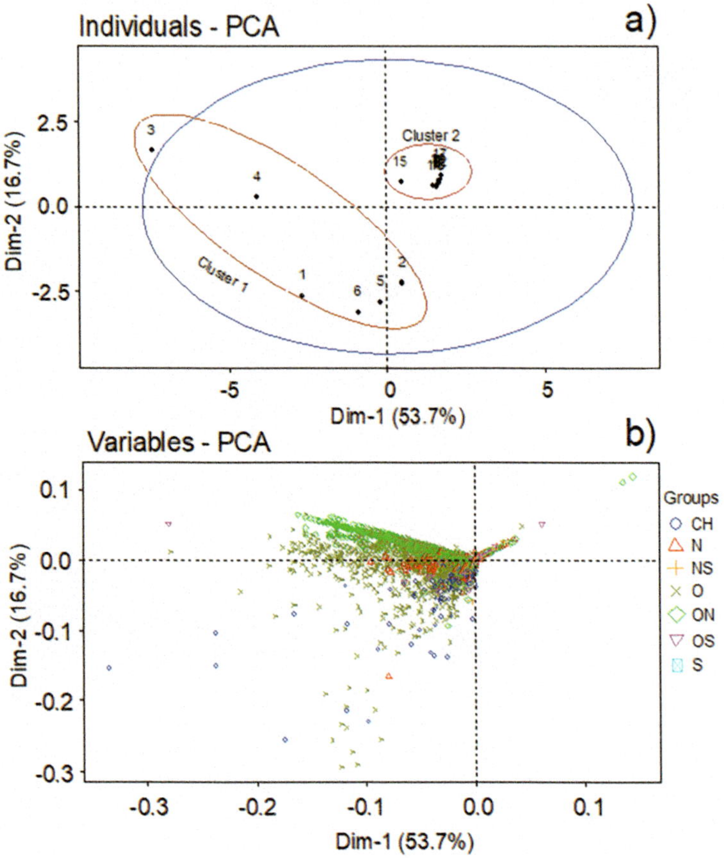

FIGURE 11.13 (a) Score; and (b) loading plots of the MS data obtained under APPI+, ESI+, and ESI-modes. Each class is represented by an individual color so that the distribution of molecular composition for soluble components can be easily recognized and evaluated.

Source: Reprinted with permission from Li et al., 2019. © John Wiley and Son.

KEYWORDS

- **atomic force microscopy**
- **electron microscopy**
- **Fourier transform infrared spectroscopy**
- **Fourier transform ion cyclotron**

- **high-resolution mass spectrometry**
- **mass spectrometry**
- **nuclear magnetic resonance**

REFERENCES

Abdelnur, P. V., Eberlin, L. S., De Sa, G. F., De Souza, V., & Eberlin, M. N., (2008). Single-shot biodiesel analysis: Nearly instantaneous typification and quality control solely by ambient mass spectrometry. *Anal. Chem., 80*, 7882–7886.

Alvarez-Ramírez, F., & Ruiz-Morales, Y., (2013). Island versus archipelago architecture for asphaltenes: Polycyclic aromatic hydrocarbon dimer theoretical studies. *Energy Fuels, 27*, 1791–1808.

Asher, S. A., & Johnson, C. R., (1984). Raman spectroscopy of a coal liquid shows that fluorescence interference is minimized with ultraviolet excitation. *Science, 225*, 311–313.

Ashida, R., Nakgawa, K., Oga, M., Nakagawa, H., & Miura, K., (2008). Fractionation of coal by use of high temperature solvent extraction technique and characterization of the fractions. *Fuel, 87*, 576–582.

Badre, S., Goncalves, C. C., Norinaga, K., Gustavson, G., & Mullins, O. C., (2006). Molecular size and weight of asphaltene and asphaltene solubility fractions from coals, crude oils and bitumen. *Fuel, 85*, 1–11.

Bae, E., Na, J. G., Chung, S. H., Kim, H. S., & Kim, S., (2010). Identification of about 30 000 chemical components in shale oils by electrospray ionization (ESI) and atmospheric pressure photoionization (APPI) coupled with 15 T Fourier transform ion cyclotron resonance mass spectrometry (FT-ICR MS) and a comparison to conventional oil. *Energy Fuels, 24*, 2563–2569.

Beckey, H. D., & Schulten, H. R., (1975). Field desorption mass spectrometry. *Angew. Chem. Int. Ed. Engl., 14*, 403–415.

Bissada, K. A., Tan, J., Szymczyk, E., Darnell, M., & Mei, M., (2016). Group-type characterization of crude oil and bitumen. Part I: Enhanced separation and quantification of saturates, aromatics, resins and asphaltenes (SARA). *Org. Geochem., 95*, 21–28.

Byrdwell, W. C., (2001). Atmospheric pressure chemical ionization mass spectrometry for analysis of lipids. *Lipids, 36*, 327–346.

Carrasco-Pancorbo, A., Nevedomskaya, E., Arthen-Engeland, T., Zey, T., Zurek, G., Baessmann, C., Deelder, A. M., & Mayboroda, O. A., (2009). Gas chromatography/atmospheric pressure chemical ionization-time of flight mass spectrometry: Analytical validation and applicability to metabolic profiling. *Anal. Chem., 81*, 10071–10079.

Carroll, D., Dzidic, I., Stillwell, R., Horning, M., & Horning, E., (1974). Subpicogram detection system for gas-phase analysis based upon atmospheric pressure ionization (API) mass spectrometry. *Anal. Chem., 46*, 706–710.

Chacón-Patiño, M. L., Rowland, S. M., & Rodgers, R. P., (2017). Advances in asphaltene petroleomics. Part 1: Asphaltenes are composed of abundant island and archipelago structural motifs. *Energy Fuels, 31*, 13509–13518.

Chacón-Patiño, M. L., Rowland, S. M., & Rodgers, R. P., (2017). Advances in asphaltene petroleomics. Part 2: Selective separation method that reveals fractions enriched in island and archipelago structural motifs by mass spectrometry. *Energy Fuels, 32*, 314–328.

Chen, Y., Mastalerz, M., & Schimmelmann, A., (2012). Characterization of chemical functional groups in macerals across different coal ranks via micro-FTIR spectroscopy. *Int. J. Coal Geol., 104*, 22–33.

Chiaberge, S., Fiorani, T., Savoini, A., Bionda, A., Ramello, S., Pastori, M., & Cesti, P., (2013). Classification of crude oil samples through statistical analysis of APPI FTICR mass spectra. *Fuel Process. Technol., 106*, 181–185.

Cho, Y., Na, J. G., Nho, N. S., Kim, S., & Kim, S., (2012). Application of saturates, aromatics, resins, and asphaltenes crude oil fractionation for detailed chemical characterization of heavy crude oils by Fourier transform ion cyclotron resonance mass spectrometry equipped with atmospheric pressure photoionization. *Energy Fuels, 26*, 2558–2565.

Christensen, G., Apelian, M. R., Hickey, K. J., & Jaffe, S. B., (1999). Future directions in modeling the FCC process: An emphasis on product quality. *Chem. Eng. Sci., 54*, 2753–2764.

Christensen, J. H., Hansen, A. B., Tomasi, G., Mortensen, J., & Andersen, O., (2004). Integrated methodology for forensic oil spill identification. *Environ. Sci. Technol., 38*, 2912–2918.

Chung, H., & Ku, M. S., (2000). Comparison of near-infrared, infrared, and Raman spectroscopy for the analysis of heavy petroleum products. *Appl. Spectrosc., 54*, 239–245.

Colati, K. A., Dalmaschio, G. P., De Castro, E. V., Gomes, A. O., Vaz, B. G., & Romão, W., (2013). Monitoring the liquid/liquid extraction of naphthenic acids in Brazilian crude oil using electrospray ionization FT-ICR mass spectrometry (ESI FT-ICR MS). *Fuel, 108*, 647–655.

Cole, D. P., Smith, E. A., Dalluge, D., Wilson, D. M., Heaton, E. A., Brown, R. C., & Lee, Y. J., (2013). Molecular characterization of nitrogen-containing species in switchgrass bio-oils at various harvest times. *Fuel, 111*, 718–726.

Corilo, Y. E., Rowland, S. M., & Rodgers, R. P., (2016). Calculation of the total sulfur content in crude oils by positive-ion atmospheric pressure photoionization Fourier transform ion cyclotron resonance mass spectrometry. *Energy Fuels, 30*, 3962–3966.

Da Silveira, G. D., Faccin, H., Claussen, L., Goularte, R. B., Do Nascimento, P. C., Bohrer, D., Cravo, M., et al., (2016). A liquid chromatography atmospheric pressure photoionization tandem mass spectrometric method for the determination of organosulfur compounds in petroleum asphalt cements. *J. Chromatogr. A, 1457*, 29–40.

Dearing, T. I., Thompson, W. J., Rechsteiner, C. E., & Marquardt, B. J., (2011). Characterization of crude oil products using data fusion of process Raman, infrared, and nuclear magnetic resonance (NMR) spectra. *Appl. Spectrosc., 65*, 181–186.

DeCanio, S. J., Nero, V. P., DeTar, M. M., & Storm, D. A., (1990). Determination of the molecular weights of the Ratawi vacuum residue fractions—a comparison of mass spectrometric and vapor phase osmometry techniques. *Fuel, 69*, 1233–1236.

Dias, H. P., Dixini, P. V., Almeida, L. C., Vanini, G., Castro, E. V., Aquije, G. M., Gomes, A. O., et al., (2015). Evidencing the crude oil corrosion by Raman spectroscopy, atomic force microscopy and electrospray ionization FT-ICR mass spectrometry. *Fuel, 139*, 328–336.

Dong, X., Wang, F., Fan, X., Zhao, Y. P., Wei, X. Y., Wang, R. Y., Ma, F. Y., et al., (2019). Evaluation of elemental composition obtained by using mass spectrometer and elemental analyzer: A case study on model compound mixtures and a coal-derived liquid. *Fuel, 245*, 392–397.

Dong, X., Zhang, Y., Milton, J., Yerabolu, R., Easterling, L., & Kenttämaa, H. I., (2019). Investigation of the relative abundances of single-core and multicore compounds in asphaltenes by using high-resolution in-source collision-activated dissociation and medium-energy collision-activated dissociation mass spectrometry with statistical considerations. *Fuel, 246*, 126–132.

Evdokimov, I. N., Eliseev, N. Y., & Akhmetov, B. R., (2003). Initial stages of asphaltene aggregation in dilute crude oil solutions: Studies of viscosity and NMR relaxation. *Fuel, 82*, 817–823.

Fan, T., Wang, J., & Buckley, J. S., (2002). Evaluating crude oils by SARA analysis. In: *SPE/DOE Improved Oil Recovery Symposium.* Society of Petroleum Engineers.

Fan, X., Jiang, J., Chen, L., Zhou, C. C., Zhu, J. L., Zhu, T. G., & Wei, X. Y., (2016). Identification of organic fluorides and distribution of organic species in an anthracite with high content of fluorine. *Fuel Process. Technol., 142*, 54–58.

Fan, X., Yu, G., Wang, M., Zhao, Y. P., Wei, X. Y., Ma, F. Y., & Zhong, M., (2018). Insight into the molecular distribution of soluble components from Dayan lignite through mass spectrometers with four ionization methods. *Fuel, 227*, 177–182.

Fan, X., Yu, Y. R., Xia, J. L., Zhao, Y. P., Cao, J. P., Lu, Y., Yue, X. M., et al., (2017). Analysis of soluble components in coals and interpretations for the complex mass spectra. *Rapid Commun. Mass Spectrom., 31*, 503–508.

Fan, X., Zhang, X. Y., Dong, X., Liao, J. J., Zhao, Y. P., Wei, X. Y., Ma, F. Y., & Nulahong, A., (2018). Structural insights of four thermal dissolution products of Dongming lignite by using in-source collision-activated dissociation mass spectrometry. *Fuel, 230*, 78–82.

Feng, M., Andrews, T., & Flores, E., (2012). Hydrocarbon Processing, In: Lee, M., (ed.), *Mass Spectrometry Handbook* (pp. 707–724). John Wiley & Sons: Hoboken.

Fenn, J. B., Mann, M., Meng, C. K., Wong, S. F., & Whitehouse, C. M., (1989). Electrospray ionization for mass spectrometry of large biomolecules. *Science, 246*, 64–71.

Fernandez-Lima, F. A., Becker, C., McKenna, A. M., Rodgers, R. P., Marshall, A. G., & Russell, D. H., (2009). Petroleum crude oil characterization by IMS-MS and FTICR MS. *Anal. Chem., 81*, 9941–9947.

Frysinger, G. S., & Gaines, R. B., (1999). Comprehensive two-dimensional gas chromatography with mass spectrometric detection (GC×GC/MS) applied to the analysis of petroleum. *J. High Resolut. Chromatogr., 22*, 251–255.

Gao, J., Borton, D. J., Owen, B. C., Jin, Z., Hurt, M., Amundson, L. M., Madden, J. T., et al., (2011). Laser-induced acoustic desorption/atmospheric pressure chemical ionization mass spectrometry. *J. Am. Soc. Mass Spectrom., 22*, 531–538.

Geng, W., Nakajima, T., Takanashi, H., & Ohki, A., (2009). Analysis of carboxyl group in coal and coal aromaticity by Fourier transform infrared (FT-IR) spectrometry. *Fuel, 88*, 139–144.

Ghosh, A. K., Srivastava, S. K., & Bagchi, S., (2007). Study of self-aggregation of coal-derived asphaltene in organic solvents: A fluorescence approach. *Fuel, 86*, 2528–2534.

Ghosh, P., & Jaffe, S. B., (2006). Detailed composition-based model for predicting the cetane number of diesel fuels. *Ind. Eng. Chem. Res., 45*, 346–351.

Ha, J., Cho, E., & Kim, S., (2017). Interpreting chemical structures of compounds in crude oil based on the tandem mass spectra of standard compounds obtained at the same normalized collision energy. *Energy Fuels, 31*, 6960–6967.

Hasan, M. U., Ali, M. F., & Bukhari, A., (1983). Structural characterization of Saudi Arabian heavy crude oil by NMR spectroscopy. *Fuel, 62*, 518–523.

Headley, J. V., Peru, K. M., & Barrow, M. P., (2016). Advances in mass spectrometric characterization of naphthenic acids fraction compounds in oil sands environmental samples and crude oil—a review. *Mass Spectrom. Rev., 35*, 311–328.

Headley, J. V., Peru, K. M., Janfada, A., Fahlman, B., Gu, C., & Hassan, S., (2011). Characterization of oil sands acids in plant tissue using orbitrap ultra-high resolution mass spectrometry with electrospray ionization. *Rapid Commun. Mass Spectrom., 25*, 459–462.

Hourani, N., Muller, H., Adam, F. M., Panda, S. K., Witt, M., Al-Hajji, A. A., & Sarathy, S. M., (2015). Structural level characterization of base oils using advanced analytical techniques. *Energy Fuels, 29*, 2962–2970.

Hsu, C. S., Dechert, G., Robbins, W., & Fukuda, E., (2000). Naphthenic acids in crude oils characterized by mass spectrometry. *Energy Fuels, 14*, 217–223.

Hsu, C. S., Hendrickson, C. L., Rodgers, R. P., McKenna, A. M., & Marshall, A. G., (2011). Petroleomics: Advanced molecular probe for petroleum heavy ends. *J. Mass Spectrom., 46*, 337–343.

Jaffe, S. B., Freund, H., & Olmstead, W. N., (2005). Extension of structure-oriented lumping to vacuum residua. *Ind. Eng. Chem. Res., 44*, 9840–9852.

Jarrell, T. M., Jin, C., Riedeman, J. S., Owen, B. C., Tan, X., Scherer, A., Tykwinski, R. R., et al., (2014). Elucidation of structural information achievable for asphaltenes via collision-activated dissociation of their molecular ions in MSn experiments: A model compound study. *Fuel, 133*, 106–114.

Jewell, D., Weber, J., Bunger, J., Plancher, H., & Latham, D., (1972). Ion-exchange, coordination, and adsorption chromatographic separation of heavy-end petroleum distillates. *Anal. Chem., 44*, 1391–1395.

Jin, C., Viidanoja, J., Li, M., Zhang, Y., Ikonen, E., Root, A., Romanczyk, M., et al., (2016). Comparison of atmospheric pressure chemical ionization and field ionization mass spectrometry for the analysis of large saturated hydrocarbons. *Anal. Chem., 88*, 10592–10598.

Jing, Z., Rodrigues, S., Strounina, E., Li, M., Wood, B., Underschultz, J. R., Esterle, J. S., & Steel, K. M., (2019). Use of FTIR, XPS, NMR to characterize oxidative effects of NaClO on coal molecular structures. *Int. J. Coal Geol., 201*, 1–13.

Jovanovic, B., (1982). Selection and evolution of industry. *Econometrica, 50*, 649–670.

Karaca, F., Morgan, T., George, A., Bull, I., Herod, A., Millan, M., & Kandiyoti, R., (2009). Molecular mass ranges of coal tar pitch fractions by mass spectrometry and size-exclusion chromatography. *Rapid Commun Mass Spectrom., 23*, 2087–2098.

Kim, E., No, M. H., Koh, J., & Kim, S., (2011). Compositional characterization of petroleum heavy oils generated from vacuum distillation and catalytic cracking by positive-mode APPI FT-ICR mass spectrometry. *Mass Spectrom. Lett., 2*, 41–44.

Kim, S., Rodgers, R. P., & Marshall, A. G., (2006). Truly "exact" mass: Elemental composition can be determined uniquely from molecular mass measurement at~ 0.1 mDa accuracy for molecules up to~ 500 Da. International *J. Mass Spectrom., 251*, 260–265.

Lai, W. C., & Song, C., (1995). Temperature-programmed retention indices for gc and gc-ms analysis of coal-and petroleum-derived liquid fuels. *Fuel, 74*, 1436–1451.

Lalli, P. M., Corilo, Y. E., Rowland, S. M., Marshall, A. G., & Rodgers, R. P., (2015). Isomeric separation and structural characterization of acids in petroleum by ion mobility mass spectrometry. *Energy Fuels, 29*, 3626–3633.

Li, B., Fan, X., Yu, Y. R., Wang, F., Li, X., Lu, Y., Wei, X. Y., et al., (2019). Insight into molecular information of Huolinguole lignite obtained by FT-ICR mass spectrometry and statistical methods. *Rapid Commun. Mass Spectrom., 33*, 1107–1113.

Liu, Z., & Phillips, J. B., (1991). Comprehensive two-dimensional gas chromatography using an on-column thermal modulator interface. *J. Chromatogr. Sci., 29*, 227–231.

Lobodin, V. V., Nyadong, L., Ruddy, B. M., Curtis, M., Jones, P. R., Rodgers, R. P., & Marshall, A. G., (2015). DART Fourier transform ion cyclotron resonance mass spectrometry for analysis of complex organic mixtures. *Int. J. Mass Spectrom., 378*, 186–192.

Lozano, D. C. P., Orrego-Ruiz, J. A., Hernández, R. C., Guerrero, J. E., & Mejía-Ospino, E., (2017). APPI (+)-FTICR mass spectrometry coupled to partial least squares with genetic algorithm variable selection for prediction of API gravity and CCR of crude oil and vacuum residues. *Fuel, 193*, 39–44.

Lu, L., Sahajwalla, V., Kong, C., & Harris, D., (2001). Quantitative X-ray diffraction analysis and its application to various coals. *Carbon, 39*, 1821–1833.

Mamin, G., Gafurov, M., Yusupov, R., Gracheva, I., Ganeeva, Y. M., Yusupova, T., & Orlinskii, S., (2016). Toward the asphaltene structure by electron paramagnetic resonance relaxation studies at high fields (3.4 T). *Energy Fuels, 30*, 6942–6946.

Mapolelo, M. M., Rodgers, R. P., Blakney, G. T., Yen, A. T., Asomaning, S., & Marshall, A. G., (2011). Characterization of naphthenic acids in crude oils and naphthenates by electrospray ionization FT-ICR mass spectrometry. *Int. J. Mass Spectrom., 300*, 149–157.

Marshall, A. G., & Rodgers, R. P., (2004). Petroleomics: The next grand challenge for chemical analysis. *Acc. Chem. Res., 37*, 53–59.

McKenna, A. M., Purcell, J. M., Rodgers, R. P., & Marshall, A. G., (2009). Identification of vanadyl porphyrins in a heavy crude oil and raw asphaltene by atmospheric pressure photoionization Fourier transform ion cyclotron resonance (FT-ICR) mass spectrometry. *Energy Fuels, 23*, 2122–2128.

Melendez, L. V., Lache, A., Orrego-Ruiz, J. A., Pachón, Z., & Mejía-Ospino, E., (2012). Prediction of the SARA analysis of Colombian crude oils using ATR–FTIR spectroscopy and chemometric methods. *J. Pet. Sci. Eng., 90*, 56–60.

Mullins, O. C., (2010). The modified yen model. *Energy Fuels, 24*, 2179–2207.

Mullins, O. C., Sabbah, H., Eyssautier, J., Pomerantz, A. E., Barré, L., Andrews, A. B., Ruiz-Morales, Y., et al., (2012). Advances in asphaltene Science and the Yen–Mullins model. *Energy Fuels, 26*, 3986–4003.

Mullins, O. C., Sheu, E. Y., Hammami, A., & Marshall, A. G., (2007). *Asphaltenes, Heavy Oils, and Petroleomics.* Springer Science & Business Media: New York.

Nyadong, L., Lai, J., Thompsen, C., LaFrancois, C. J., Cai, X., Song, C., Wang, J., & Wang, W., (2017). High-field orbitrap mass spectrometry and tandem mass spectrometry for molecular characterization of asphaltenes. *Energy Fuels, 32*, 294–305.

Owen, B. C., Gao, J., Borton, D. J., Amundson, L. M., Archibold, E. F., Tan, X., Azyat, K., et al., (2011). Carbon disulfide reagent allows the characterization of nonpolar analytes by atmospheric pressure chemical ionization mass spectrometry. *Rapid Commun. Mass Spectrom., 25*, 1924–1928.

Pérez-Hernández, R., Mendoza-Anaya, D., Mondragon-Galicia, G., Espinosa, M., Rodrıguez-Lugo, V., Lozada, M., & Arenas-Alatorre, J., (2003). Microstructural study of asphaltene precipitated with methylene chloride and n-hexane. *Fuel, 82*, 977–982.

Podgorski, D. C., Corilo, Y. E., Nyadong, L., Lobodin, V. V., Bythell, B. J., Robbins, W. K., McKenna, A. M., et al., (2013). Heavy petroleum composition. 5. Compositional and structural continuum of petroleum revealed. *Energy Fuels, 27*, 1268–1276.

Portolés, T., Cherta, L., Beltran, J., & Hernández, F., (2012). Improved gas chromatography-tandem mass spectrometry determination of pesticide residues making use of atmospheric pressure chemical ionization. *J. Chromatogr. A, 1260*, 183–192.

Portolés, T., Sancho, J., Hernández, F., Newton, A., & Hancock, P., (2010). Potential of atmospheric pressure chemical ionization source in GC-QTOF MS for pesticide residue analysis. *J. Mass Spectrom, 45*, 926–936.

Purcell, J. M., Hendrickson, C. L., Rodgers, R. P., & Marshall, A. G., (2006). Atmospheric pressure photoionization Fourier transform ion cyclotron resonance mass spectrometry for complex mixture analysis. *Anal. Chem., 78*, 5906–5912.

Putman, J. C., Rowland, S. M., Podgorski, D. C., Robbins, W. K., & Rodgers, R. P., (2017). Dual-column aromatic ring class separation with improved universal detection across mobile-phase gradients via eluate dilution. *Energy Fuels, 31*, 12064–12071.

Qian, K., Edwards, K. E., Mennito, A. S., Freund, H., Saeger, R. B., Hickey, K. J., Francisco, M. A., et al., (2012). Determination of structural building blocks in heavy petroleum systems by collision-induced dissociation Fourier transform ion cyclotron resonance mass spectrometry. *Anal. Chem., 84*, 4544–4551.

Qian, K., Robbins, W. K., Hughey, C. A., Cooper, H. J., Rodgers, R. P., & Marshall, A. G., (2001). Resolution and identification of elemental compositions for more than 3,000 crude acids in heavy petroleum by negative-ion micro electrospray high-field Fourier transform ion cyclotron resonance mass spectrometry. *Energy Fuels, 15*, 1505–1511.

Quann, R. J., & Jaffe, S. B., (1992). Structure-oriented lumping: Describing the chemistry of complex hydrocarbon mixtures. *Ind. Eng. Chem. Res., 31*, 2483–2497.

Rana, M. S., Samano, V., Ancheyta, J., & Diaz, J., (2007). A review of recent advances on process technologies for upgrading of heavy oils and residua. *Fuel, 86*, 1216–1231.

Rausa, R., Calemma, V., Ghelli, S., & Girardi, E., (1989). Study of low temperature coal oxidation by 13C CP/MAS NMR. *Fuel, 68*, 1168–1172.

Robb, D. B., Covey, T. R., & Bruins, A. P., (2000). Atmospheric pressure photoionization: An ionization method for liquid chromatography-mass spectrometry. *Anal. Chem., 72*, 3653–3659.

Robson, W. J., Sutton, P. A., McCormack, P., Chilcott, N. P., & Rowland, S. J., (2017). Class type separation of the polar and apolar components of petroleum. *Anal. Chem., 89*, 2919–2927.

Rodgers, R. P., & McKenna, A. M., (2011). Petroleum analysis. *Anal. Chem., 83*, 4665–4687.

Rodgers, R. P., Schaub, T. M., & Marshall, A. G., (2005). *Petroleomics: MS Returns to Its Roots*. ACS Publications.

Sabbah, H., Morrow, A. L., Pomerantz, A. E., & Zare, R. N., (2011). Evidence for island structures as the dominant architecture of asphaltenes. *Energy Fuels, 25*, 1597–1604.

Santos, J. M., Wisniewski, A., Eberlin, M. N., & Schrader, W., (2018). Comparing crude oils with different API gravities on a molecular level using mass spectrometric analysis. Part 1: Whole crude oil. *Energies, 11*, 1–11.

Santos, J., Vetere, A., Wisniewski, A., Eberlin, M., & Schrader, W., (2018). Comparing crude oils with different API gravities on a molecular level using mass spectrometric analysis. Part 2: Resins and asphaltenes. *Energies, 11*, 2767.

Schaub, T. M., Hendrickson, C. L., Qian, K., Quinn, J. P., & Marshall, A. G., (2003). High-resolution field desorption/ionization Fourier transform ion cyclotron resonance mass analysis of nonpolar molecules. *Anal. Chem., 75*, 2172–2176.

Schaub, T. M., Hendrickson, C. L., Quinn, J. P., Rodgers, R. P., & Marshall, A. G., (2005). Instrumentation and method for ultrahigh-resolution field desorption ionization Fourier transform ion cyclotron resonance mass spectrometry of nonpolar species. *Anal. Chem., 77*, 1317–1324.

Schuler, B., Fatayer, S., Meyer, G., Rogel, E., Moir, M., Zhang, Y., Harper, M. R., et al., (2017). Heavy oil-based mixtures of different origins and treatments studied by atomic force microscopy. *Energy Fuels, 31*, 6856–6861.

Schuler, B., Meyer, G., Peña, D., Mullins, O. C., & Gross, L., (2015). Unraveling the molecular structures of asphaltenes by atomic force microscopy. *J. Am. Chem. Soc., 137*, 9870–9876.

Stenson, A. C., Marshall, A. G., & Cooper, W. T., (2003). Exact masses and chemical formulas of individual Suwannee River fulvic acids from ultrahigh resolution electrospray ionization Fourier transform ion cyclotron resonance mass spectra. *Anal. Chem., 75*, 1275–1284.

Strausz, O. P., Mojelsky, T. W., & Lown, E. M., (1992). The molecular structure of asphaltene: An unfolding story. *Fuel, 71*, 1355–1363.

Strausz, O. P., Peng, P., & Murgich, J., (2002). About the colloidal nature of asphaltenes and the MW of covalent monomeric units. *Energy Fuels, 16*, 809–822.

Sun, Z. Q., Ma, F. Y., Liao, J., Liu, J. M., Zhong, M., & Kong, L. T., (2017). Effects of mechanical activation on the co-liquefaction of xigou sub-bituminous coal from Xinjiang and Karamay petroleum vacuum residues. *Fuel Process. Technol., 161*, 139–144.

Takanohashi, T., & Kawashima, H., (2002). Construction of a model structure for upper freeport coal using C-13 NMR chemical shift calculations. *Energy Fuels, 16*, 379–387.

Tanaka, K., Waki, H., Ido, Y., Akita, S., Yoshida, Y., Yoshida, T., & Matsuo, T., (1988). Protein and polymer analyses up to m/z 100 000 by laser ionization time-of-flight mass spectrometry. *Rapid Commun Mass Spectrom., 2*, 151–153.

Teng, S. T., Williams, A. D., & Urdal, K., (1994). Detailed hydrocarbon analysis of gasoline by GC-MS (SI-PIONA). *J. High Resolut. Chromatogr., 17*, 469–475.

Tose, L. V., Cardoso, F. M., Fleming, F. P., Vicente, M. A., Silva, S. R., Aquije, G. M., Vaz, B. G., & Romão, W., (2015). Analyzes of hydrocarbons by atmosphere pressure chemical ionization FT-ICR mass spectrometry using isooctane as ionizing reagent. *Fuel, 153*, 346–354.

Van, G. M., Swennen, R., & David, P., (2001). Quantitative coal characterization by means of microfocus X-ray computer tomography, color image analysis and back-scattered scanning electron microscopy. *Int. J. Coal Geol., 46*, 11–25.

Vetere, A., & Schrader, W., (2017). Mass spectrometric coverage of complex mixtures: Exploring the carbon space of crude oil. *Chemistry Select, 2*, 849–853.

Wang, F. C. Y., Qian, K., & Green, L. A., (2005). GC×MS of diesel: A two-dimensional separation approach. *Anal. Chem., 77*, 2777–2785.

Wang, F. C. Y., Robbins, W. K., Di Sanzo, F. P., & McElroy, F. C., (2003). Speciation of sulfur-containing compounds in diesel by comprehensive two-dimensional gas chromatography. *J. Chromatogr. Sci., 41*, 519–523.

Wang, F., Fan, X., Xia, J. L., Wei, X. Y., Yu, Y. R., Zhao, Y. P., Cao, J. P., et al., (2018). Insight into the structural features of low-rank coals using comprehensive two-dimensional gas chromatography/time-of-flight mass spectrometry. *Fuel, 212*, 293–301.

Wang, M., Fan, X., Wei, X. Y., Cao, J. P., Zhao, Y. P., Wang, S. Z., Wang, C. F., & Wang, R. Y., (2016). Characterization of the oxidation products of Shengli lignite using mass spectrometers with "hard", "soft" and ambient ion sources. *Fuel, 183*, 115–122.

Weldegergis, B. T., De Villiers, A., & Crouch, A. M., (2011). Chemometric investigation of the volatile content of young South African wines. *Food Chem., 128*, 1100–1109.

Wittrig, A. M., Fredriksen, T. R., Qian, K., Clingenpeel, A. C., & Harper, M. R., (2017). Single Dalton collision-induced dissociation for petroleum structure characterization. *Energy Fuels, 31*, 13338–13344.

Wu, C., Qian, K., Walters, C. C., & Mennito, A., (2015). Application of atmospheric pressure ionization techniques and tandem mass spectrometry for the characterization of petroleum components. *Int. J. Mass Spectrom., 377*, 728–735.

Xia, J. L., Fan, X., You, C. Y., Wei, X. Y., Zhao, Y. P., & Cao, J. P., (2016). Sequential ultrasonic extraction of a Chinese coal and characterization of nitrogen-containing compounds in the extracts using high-performance liquid chromatography with mass spectrometry. *J. Sep. Sci., 39*, 2491–2498.

Yao, Z. S., Wei, X. Y., Lv, J., Liu, F. J., Huang, Y. G., Xu, J. J., Chen, F. J., et al., (2010). Oxidation of Shenfu coal with RuO_4 and NaOCl. *Energy Fuels, 24*, 1801–1808.

Ye, R., Xiang, C., Lin, J., Peng, Z., Huang, K., Yan, Z., Cook, N. P., et al., (2013). Coal as an abundant source of graphene quantum dots. *Nat. Commun., 4*, 2943.

Yen, T. F., Erdman, J. G., & Pollack, S. S., (1961). Investigation of the structure of petroleum asphaltenes by X-ray diffraction. *Anal. Chem., 33*, 1587–1594.

Yerabolu, R., Kotha, R. R., Niyonsaba, E., Dong, X., Manheim, J. M., Kong, J., Riedeman, J. S., et al., (2018). Molecular profiling of crude oil by using distillation precipitation fractionation mass spectrometry (DPF-MS). *Fuel, 234*, 492–501.

Yu, Y. R., Fan, X., Chen, L., Dong, X. M., Zhao, Y. P., Li, B., Wei, X. Y., et al., (2019). Mass spectrometric evaluation of the soluble species of Shengli lignite using cluster analysis methods. *Fuel*, 236, 1037–1042.

Zubkova, V., (2011). Chromatographic methods and techniques used in studies of coals, their progenitors and coal-derived materials. *Anal. BioAnal. Chem., 399*, 3193–3209.

Index

G

K

L

M